What Einstein Did Exactly

Sandeep Chavan

Published by Gyrus Vision, 2026.

WHAT EINSTEIN DID EXACTLY

First edition. January 2, 2026.

ISBN: 979-8233529870

Written by Sandeep Chavan.

Table of Contents

To all who are willing to pause between certainty and curiosity— and listen to what understanding has not yet said.

And to Einstein

"The most incomprehensible thing about the universe is that it is comprehensible."

— Albert Einstein

Opening Note

This book is not written to praise Einstein, nor is it written in opposition to him. It is written to place his work exactly where it belongs.

Over time, Einstein has become more than a physicist. He has become a symbol, a reference point, sometimes even a shield. His name is often used to end conversations rather than begin them. That was never his intention. Einstein did not ask to be followed. He asked to be understood.

What he gave the world was not a picture of the universe, but a precise way to describe how the universe responds when energy and matter are present. His achievement was not ontological. It was relational. He did not claim to tell us what reality is made of. He showed us how consequences unfold once certain conditions exist. That distinction matters more today than it did a century ago.

The success of his equations created a quiet illusion of completeness. Because the predictions worked, the language began to harden into belief. Geometry, which began as a tool, slowly acquired the weight of substance. Curvature, which described behavior, began to be treated as an actor. Expansion, which interpreted observation, became a story about reality itself. None of this was dishonest. It was the natural outcome of a theory that worked too well.

Einstein himself remained uneasy with this closure. His later years were not a retreat into obsession, as they are often portrayed, but a refusal to accept that description alone was enough. He sensed that something essential was missing, even if he could not yet name it. He searched not to overthrow his own work, but to understand what made it possible.

This book takes that unease seriously.

It does not argue that Einstein was wrong. It argues that he stopped at the correct place. It does not attempt to replace his framework, nor does it elevate an alternative by diminishing his contribution. Instead, it asks a simple and overdue question: what exactly did Einstein do—and what did he deliberately leave untouched?

The answer matters because modern physics increasingly struggles not with calculation, but with interpretation. We predict outcomes with extraordinary accuracy, yet we remain uncertain about what produces them. We accumulate constants, corrections, and unseen components, not because the data demands confusion, but because our language has outpaced our physical understanding.

This book proposes that Einstein gave us the most refined language of consequences humanity has ever known, but that language was never meant to stand alone. It was a bridge, not a destination. Crossing that bridge does not diminish what lies behind it. It clarifies why it was built.

What follows is a careful examination of Einstein's work, his restraint, and the misinterpretations that grew around both. It is also an attempt to complete an unfinished sentence—not by rewriting history, but by continuing it with the same discipline Einstein himself practiced.

This is not a book of admiration or rebellion. It is a book of placement.

Preface

This book began with a discomfort that grew slowly rather than dramatically. It did not arise from disagreement with established physics, nor from dissatisfaction with its predictive power. On the contrary, it emerged from a deep respect for how reliably modern physics works—and from an equally deep unease about how rarely we pause to ask *what exactly it is doing* when it works.

For more than a century, physics has advanced by refining its ability to predict consequences. We have learned to calculate how light bends, how clocks drift, how galaxies rotate, and how the universe appears to evolve when observed across immense distances. These achievements are real, repeatable, and experimentally validated. Yet alongside this success runs a quieter question, one that is often postponed rather than addressed: do our descriptions explain reality, or do they merely organize its responses?

At the center of this tension stands **Albert Einstein**. His work reshaped physics not by adding new objects to the universe, but by changing the way we speak about it. He replaced force with geometry, motion with relation, and absolute structures with contextual behavior. In doing so, he gave us an extraordinarily precise language for consequences—how matter and energy affect their surroundings and how those surroundings, in turn, guide motion.

What Einstein did *not* do is equally important. He did not define what space is made of. He did not claim that geometry itself is substance. He did not assert that the universe expands because space possesses agency. These interpretations emerged later, gradually, and often unintentionally, as the success of his equations encouraged the belief that the language itself must be the reality it described.

This book is written to slow that drift.

It does not seek to correct Einstein, nor to reinterpret his equations. It seeks to clarify the boundary he drew and that later generations quietly crossed. The goal is not to dismantle modern physics, but to understand why it increasingly relies on placeholders—dark components, adjustable parameters, and interpretive patches—despite its mathematical elegance.

The central claim of this book is simple, though its implications are not: Einstein modeled *how consequences occur*, not *what produces them*. His equations work because nature is consistent in its responses, even when the underlying mechanism remains unnamed. Over time, the unnamed was mistaken for unnecessary, and description was mistaken for explanation.

This confusion has consequences of its own. When geometry is treated as substance, questions about physical medium are deferred. When expansion is treated as fact rather than interpretation, observational tension becomes paradox. When the vacuum is treated as nothing, it is asked to explain everything. None of these moves are illogical. They are the natural outcomes of a framework that achieved extraordinary success without ever claiming completeness.

The Universal Energy Dynamics (UED) framework enters this discussion not as a challenger to Einstein, but as a response to the space he intentionally left open. It proposes that the universe is not defined by expansion or contraction, but by equilibrium-seeking dynamics within a structured vacuum medium. In this view, gravity, redshift, and cosmic behavior are consequences of energy redistribution, not properties of geometry acting on its own.

This book does not present UED as a replacement theory. It presents it as a completion of an unfinished sentence. Einstein gave us the grammar of consequences. UED asks what that grammar is written upon.

The chapters that follow are written deliberately without mathematical density. This is not a concession to simplicity, but a recognition of where the problem now lies. The present crisis in cosmology is not computational. It is interpretive. Adding equations before resolving meaning would only deepen the confusion this book aims to clarify.

The structure of the book follows a continuous dialogue rather than a sequence of arguments. Each chapter builds upon the previous one without attempting to persuade or provoke. The tone is measured because the subject demands maturity. Einstein does not need to be defended, nor does modern physics need to be dismantled. What is needed is careful placement—of ideas, assumptions, and limits.

Readers are not expected to agree with every conclusion offered here. They are asked only to follow the distinction the book maintains throughout: the difference between describing consequences and explaining causes. That distinction, once seen clearly, alters how familiar theories are understood and how unresolved tensions are interpreted.

This is not a book about overthrowing a legacy. It is a book about understanding it precisely enough to move forward without distortion.

If the reader finishes this book with greater respect for Einstein's restraint, greater clarity about what modern physics does well, and a renewed willingness to ask foundational questions without impatience, then it will have served its purpose.

How to Read This Book

This book is not meant to be read quickly, nor is it meant to convince. It is designed to be *encountered* rather than consumed. The ideas unfold gradually, and their value lies less in agreement than in reflection.

You do not need to read it linearly, though doing so will reveal an intentional progression. Each chapter builds upon the previous one, not by adding complexity, but by shifting perspective. The early sections establish orientation; the later ones deepen it. Together, they form a continuous inquiry rather than a sequence of arguments.

This is not a textbook, and it does not offer formulas, proofs, or definitive answers. It also does not ask you to accept a new doctrine. Instead, it invites you to slow down and examine how you think about explanation, structure, and meaning—particularly in relation to science and reality.

Some readers may find it helpful to read slowly, even revisiting certain sections. The ideas are not meant to be absorbed in a single pass. They are meant to settle, to provoke quiet reconsideration, and to be revisited from different angles. If a passage feels unfamiliar or challenging, that is intentional; it often signals a shift in perspective rather than a gap in understanding.

You may also find that certain sections resonate more strongly than others depending on your background. Readers with scientific training may connect first with the conceptual analysis. Those inclined toward philosophy may find resonance in the reflective passages. Both approaches are valid, and neither is required to grasp the whole.

This book does not ask for agreement. It asks for attention. It does not present conclusions to be accepted, but ideas to be explored. Its purpose is not to persuade, but to open space for thought.

If you finish the book with more questions than you began with, it has succeeded. If it encourages you to pause, reconsider, or see familiar ideas in a new light, it has done its work.

Read it not as a destination, but as a companion to inquiry—one that walks beside you rather than ahead of you.

Part I — The Boundary Of Understanding

Establishing what science explains—and what it deliberately leaves untouched.

Chapter 1 — The Boundary Einstein Drew

"It is the theory which decides what we can observe." — **Albert Einstein**

The Boundary That Was Never Crossed

At some point in the last century, a quiet substitution took place. It was not announced, debated, or even noticed clearly. A way of *describing* reality slowly began to stand in for reality itself.

Einstein never asked for this substitution.

What he offered physics was a method of extraordinary restraint. He showed how matter and energy influence their surroundings and how those surroundings, in turn, shape motion. He provided a language that captured consequences with remarkable accuracy. But he stopped there. He did not claim that the language was the thing being spoken about.

That stopping point matters.

Modern physics often speaks as if Einstein told us what the universe *is*. In truth, he told us how the universe *behaves* once certain conditions are present. Curvature, fields, and spacetime entered physics not as substances, but as relationships. They were answers to "what happens?" not to "what exists?"

The boundary between those two questions is the boundary Einstein drew deliberately. On one side lies description:

measurable, testable, predictive. On the other lies definition: claims about what reality is made of. Einstein advanced science by staying rigorously on the first side, even when success tempted him to cross over.

This chapter begins by restoring that boundary.

Not to diminish Einstein's work, but to place it precisely where it belongs. Because only when that line is seen clearly can we understand both why modern physics works so well—and why it increasingly struggles to explain what its own success means.

Before we ask what the universe is, we must first understand what Einstein chose to describe—and why he chose not to go further.

1.1 — The Difference Between Knowing and Naming

There is a quiet difference between knowing how something behaves and knowing what it is. Most of the time, this difference does not trouble us. In everyday life, we cross bridges without knowing the chemistry of steel, use electricity without understanding electrons, and trust clocks without questioning time itself. Function precedes definition, and usefulness often replaces explanation. Physics, despite its reputation for depth, operates in much the same way.

To know, in the scientific sense, is to predict. If a model tells us what will happen under given conditions, and those predictions consistently match observation, we call that knowledge. Naming, on the other hand, is a different act. Naming asserts what something fundamentally is, what kind of entity exists beneath the behavior we observe. One can know without naming, and historically, physics has done so with remarkable success.

This distinction is not a philosophical indulgence; it is a practical working principle. Physics emerged not as a catalog of substances, but as a discipline of relationships. It asked how quantities change, how systems respond, how outcomes depend on conditions. When these relationships were captured reliably, the job was considered done. Whether the underlying reality was fully understood was often treated as a separate, sometimes unnecessary, concern.

This approach explains why physics can function, and often does function brilliantly, without ontology. Newton never explained what gravity is. He did not describe a medium, a mechanism, or a cause. He provided a rule that told us how objects accelerate in one another's presence. That rule worked so well that the absence of explanation was tolerated for centuries. The success of prediction made the lack of naming feel irrelevant.

The same pattern repeated at deeper levels. Electricity worked long before electrons were identified. Heat was modeled effectively before atoms were accepted. Even today, quantum mechanics predicts outcomes with astonishing accuracy while remaining ambiguous about what its mathematical entities represent physically. Wavefunctions are calculated, probabilities assigned, experiments confirmed—yet what the wavefunction *is* remains unsettled. Physics continues, largely unimpeded.

This is not a failure of the discipline. It is a feature of its method.

The danger appears only when the line between knowing and naming quietly dissolves. When a description becomes reliable enough, it begins to feel real. When equations predict perfectly, the temptation arises to treat their elements as entities rather than as tools. What began as a language of behavior slowly acquires the status of substance.

This is where Einstein's contribution must be understood carefully.

Einstein inherited a field—gravity—that was known only through its effects. Objects fell, planets moved, light bent. Newton's equations described these behaviors without naming their cause. Einstein advanced this tradition rather than breaking it. He did not ask what gravity is made of. He asked how the presence of mass and energy changes the way other things behave. His answer was geometry—not as material, but as relationship.

Spacetime curvature was a way to encode how motion responds to energy. It was a precise grammar for consequences. It told us how clocks tick differently, how paths bend, how trajectories evolve. At no point did Einstein insist that spacetime is a substance that acts on its own. He provided a framework that worked because it predicted responses accurately, not because it defined reality completely.

This restraint is easy to miss precisely because the results were so powerful. When predictions succeed across experiments, scales, and decades, it becomes psychologically difficult to remember that the model is a description. Curvature starts to feel like a thing. Expansion begins to sound like an action. The language of consequence is quietly promoted to the language of being.

Yet physics itself offers a warning here. Knowing how something behaves does not automatically tell us what it is. A thermometer measures temperature, not heat's essence. A map guides travel without being the territory. Equations can encode relationships without exhausting reality. Confusing these levels does not usually break calculations—but it distorts interpretation.

Einstein understood this risk intuitively. He was cautious with words, reluctant to declare final answers, and uncomfortable when his work was treated as complete. His later dissatisfaction was not a rejection of his earlier success, but an acknowledgment of its limits. He sensed that describing responses, no matter how accurately, was not the same as identifying the responder.

Modern physics, however, often speaks as if this distinction no longer matters. Because the models work, they are treated as reality. Because geometry predicts behavior, geometry is assumed to be fundamental. Because expansion fits data, expansion becomes a fact rather than an interpretation. The difference between knowing and naming fades into the background.

This book insists on bringing it back to the foreground.

Understanding this distinction does not weaken science. It strengthens it. It reminds us that prediction is not explanation, that description is not ontology, and that success does not equal completion. Physics can function without naming reality—but it cannot indefinitely avoid the consequences of mistaking its language for substance.

Before we ask what the universe is, we must recognize what our theories have actually given us. They have taught us how the universe responds, not what it is made of. That realization is not a criticism. It is a clarification.

Einstein drew that boundary with care. The chapters ahead begin by standing exactly where he stood—on the side of consequences—before asking, with equal care, what lies beyond that line.

1.2 — Why the Universe Was Never Defined

Einstein never defined what the universe is, not because he lacked imagination or courage, but because he understood the cost of doing so prematurely. This restraint is often mistaken for incompleteness or hesitation. In reality, it was a deliberate methodological choice—one that protected physics from collapsing into speculation while allowing it to advance with precision.

To define what the universe *is* requires more than describing how it behaves. It requires an ontological commitment: a claim about substance, structure, and ultimate composition. Einstein recognized that physics, at the stage it had reached in the early twentieth century, was not equipped to make such claims responsibly. The tools available—mathematics, observation, experiment—were exquisitely suited to modeling relationships and outcomes, but poorly suited to declaring what reality is made of beneath those outcomes.

This was not philosophical caution in the abstract. It was grounded in experience. Classical physics had already demonstrated how dangerous premature naming could be. The ether, once proposed as the medium of light, was defined too confidently and too early. It solved certain conceptual problems but failed experimental scrutiny. Einstein learned from this episode. He did not want spacetime to become another ether—an imagined substance propped up by necessity rather than evidence.

Instead, he chose a safer path: describe the universe through its responses. Ask how clocks behave in motion, how rods change under acceleration, how light propagates near mass. These were questions that could be answered operationally. They did not require metaphysical declarations. They required consistency, testability, and restraint.

This is why Einstein's universe is always encountered indirectly. It appears through time dilation, length contraction, gravitational redshift, orbital precession, and light bending. Nowhere does he pause to say what space *is*, or what time *consists of*. He treats them as relational quantities, defined by measurement and interaction, not by substance. The universe, in his work, is never an object to be described in isolation. It is a web of relations that reveals itself only through consequences.

This choice often frustrates later readers. There is a temptation to ask: if spacetime curves, what is it made of? If geometry changes, what is changing? If the universe expands, what exactly is expanding? Einstein resisted answering these questions not because they were meaningless, but because answering them without a physical basis would turn description into belief.

His caution becomes clearer when we examine his attitude toward completion. Despite the overwhelming success of General Relativity, Einstein never treated it as the final word. He did not celebrate closure. He searched for deeper unity. His pursuit of a unified field theory was not an attempt to replace relativity, but to understand what made it work so reliably. That search itself is evidence that he did not believe he had defined reality.

Einstein understood something subtle but profound: defining the universe is irreversible. Once a definition is accepted, it shapes interpretation, constrains imagination, and resists revision. Descriptions can be updated. Definitions, once institutionalized, become dogma. By refusing to define the universe, Einstein preserved openness. He left room for future physics to discover what lay beneath the relationships he had mapped so accurately.

This restraint mattered because physics does not progress in straight lines. It advances through layers. First, behavior is observed. Then patterns are described. Only later, sometimes much later, are physical mechanisms identified. Skipping steps feels efficient, but it often leads to confusion. Einstein refused to skip. He stabilized the descriptive layer before allowing deeper claims to emerge.

The scientific community, however, gradually moved beyond this restraint. As Einstein's equations succeeded across domains, the absence of an explicit definition began to feel like a gap that needed filling. Geometry quietly shifted from language to substance. Expansion shifted from interpretation to fact. The universe, which Einstein had left undefined, was slowly given a shape, a history, and a destiny.

This shift was understandable. Humans are uncomfortable with open-endedness. A universe that behaves without being defined feels incomplete. But the discomfort did not originate in Einstein's work. It originated in our desire for closure. The decision to define the universe came later—and it came without Einstein's methodological caution.

Recognizing this matters today because many of the tensions in modern cosmology trace back to that moment. When the universe is defined too early, observations that do not fit the definition become problems rather than clues. Dark components are introduced not because they are observed directly, but because the definition demands them. Expansion rates are reconciled not by rethinking assumptions, but by adjusting parameters.

Einstein avoided this trap by refusing to define what could not yet be grounded physically. He trusted that accurate description would eventually lead to deeper understanding, but only if the distinction between behavior and being was preserved.

This book does not argue that Einstein should have defined the universe. It argues that we should remember why he did not. His restraint was not a limitation of vision. It was a discipline of honesty. He defined what could be measured and left open what could not. That choice allowed physics to flourish.

The chapters that follow take that restraint seriously. They do not rush to define the universe either. Instead, they ask whether the time has come to name what Einstein deliberately left unnamed—and whether doing so now can be done without repeating the mistakes he worked so carefully to avoid.

1.3 — Consequences as the Safest Territory

Physics learned early that the safest place to stand is not at the center of reality, but at its edges—where effects appear, where measurements can be made, and where disagreement can be settled by observation rather than belief. This is why consequences became the preferred territory of the discipline. They offered stability in a landscape where causes were often hidden, inaccessible, or poorly understood.

To work with consequences is to work with what can be verified. When an object falls, when light bends, when a clock slows, something has happened that can be recorded and compared. The cause may be debated, postponed, or left unnamed, but the effect itself is not negotiable. Physics, seeking reliability over completeness, chose to anchor itself there.

This choice was not accidental. It emerged from hard-earned lessons. Every time physics tried to move too quickly into defining underlying reality, it paid a price. Speculative entities proliferated, debates hardened, and experimental clarity was lost. By contrast, when physics restricted itself to describing how systems respond under given conditions, progress resumed. Consequences could be tested. Definitions could not.

Newton's gravity is a textbook example. He described how masses attract without ever explaining why they do so. The lack of mechanism did not weaken the theory. On the contrary, it allowed it to function universally. Objects behaved as predicted regardless of whether gravity was imagined as action-at-a-distance, divine influence, or something else entirely. Consequence-based description proved robust precisely because it was agnostic about cause.

This pattern repeated with remarkable consistency. Thermodynamics advanced without knowing what heat was. Electromagnetism flourished before electrons were discovered. Quantum mechanics became the most precise predictive framework in history while remaining conceptually unsettled about what its mathematical objects represent. Each time, physics progressed by securing the behavior first and postponing the question of essence.

Einstein understood this pattern deeply. When he reformulated gravity, he did not attempt to identify a force carrier, medium, or substance. He asked a different question: given the presence of mass and energy, how do other objects behave? How do paths change? How do clocks tick? How does light travel? By framing gravity as a pattern of response rather than an interaction between things, he placed his theory squarely within the safest territory physics knew.

The result was General Relativity—a framework that predicted consequences with extraordinary accuracy. Light bending, gravitational redshift, time dilation, orbital precession—each confirmed the theory's power. None required knowing what spacetime is made of. None required assigning agency to geometry itself. The theory succeeded because it remained focused on outcomes.

This focus protected physics from premature speculation. By refusing to define the underlying substance of spacetime, Einstein avoided repeating the error of the ether. He allowed geometry to function as a language for relationships rather than as a physical claim. The theory could be tested without committing to metaphysical positions that might later prove untenable.

The protection consequences offer goes beyond caution; it enables continuity. When theories are grounded in observable behavior, they remain usable even as interpretations evolve. Newtonian mechanics still works within its domain despite being conceptually superseded. Maxwell's equations remain valid regardless of how we understand fields at a deeper level. Einstein's equations continue to predict correctly even as debates about spacetime's nature persist.

This continuity is possible only because the theories were built on consequences rather than on rigid definitions of reality. A theory that defines too much risks collapse when its definitions fail. A theory that describes behavior can survive reinterpretation.

However, safety has a side effect. When physics stays too long in the territory of consequences, it can forget that it is standing there by choice. The success of prediction begins to feel like explanation. The boundary between describing what happens and explaining why it happens becomes blurred. What was once a deliberate restraint slowly turns into an unexamined assumption.

This is where modern physics finds itself today. Its models predict with precision, yet their interpretations increasingly strain coherence. Dark components are introduced to preserve descriptive success. Expansion is treated as literal because it fits observed redshift. Geometry is spoken of as if it acts, stretches, or evolves on its own. None of these moves break calculations. They preserve consequence-matching. But they shift the theory away from the discipline that once protected it.

Einstein never made that shift himself. He remained aware that describing responses was not the same as identifying causes. His discomfort with closure, his continued search for unification, and his reluctance to assign substance to geometry all point to an awareness of the limits of consequence-based description.

The scientific community, however, gradually crossed into a different posture. Theories that worked were treated as explanations. Language hardened into ontology. The safest territory became the only territory, and physics stopped asking whether it had reached the edge of what description alone could provide.

Understanding why consequences were the safest territory helps explain both the success and the current tension in physics. Success came from restraint. Tension arises when restraint is mistaken for completion. When unanswered questions are treated as settled, anomalies appear as problems rather than as signals.

This book does not argue that physics should abandon consequences. It argues that we should remember why we chose them in the first place. Consequences are safe because they are observable. They are powerful because they are testable. But they are not the end of inquiry. They are the foundation upon which deeper understanding can be built—if we are willing to step beyond them carefully.

Einstein stood firmly on that foundation without claiming it was the whole structure. Recognizing this allows us to appreciate his achievement without inheriting his silence as a permanent rule. The safest territory was never meant to be the final destination. It was a place to stand while preparing for the next step.

1.4 — The Silent Agreement Science Made

At some point, without announcement or debate, the scientific community reached a quiet understanding. It was never written into textbooks as a principle, never voted on, never defended openly. Yet it shaped a century of research as decisively as any formal doctrine. The agreement was simple: physics would concern itself with what can be calculated and tested, and it would refrain—politely but firmly—from declaring what reality ultimately is.

This agreement was not cynical. It was practical.

After the upheavals of the early twentieth century, physics found itself in unfamiliar territory. Classical intuitions had failed. Space and time were no longer absolute. Particles behaved like waves, and waves behaved like probabilities. In such an environment, insisting on ontological clarity would have stalled progress. The discipline chose survival over certainty, prediction over metaphysics. It learned to live comfortably with unanswered questions, as long as the answers it *did* give matched experiment.

No one embodied this posture more convincingly than **Einstein**. His work demonstrated that one could achieve extraordinary explanatory reach without ever crossing into ontological declaration. He showed that physics could reorganize its foundations—space, time, gravity—while still avoiding claims about substance. The community absorbed this lesson quickly, even if it never articulated it explicitly.

What followed was not a conspiracy or a suppression of inquiry, but a shared discipline. Papers focused on derivations, predictions, and consistency. Questions about what spacetime *is*, what fields *are*, or what the vacuum *contains* were treated as secondary, sometimes even as distractions. They were not forbidden, but they were gently sidelined. Career incentives, funding structures, and publication norms all reinforced the same message: work where results can be checked.

This silent agreement proved remarkably effective. Physics advanced at an unprecedented pace. Technologies emerged. Predictions tightened. Experimental confirmations accumulated. The community learned that it could move forward without settling foundational questions—and that doing so was often the fastest path to success.

Over time, this practice hardened into habit.

Students were trained to manipulate equations long before asking what those equations meant physically. Conceptual discomfort was reframed as philosophical curiosity, something to be entertained after calculations were done—if at all. When models worked, they were treated as sufficient. When they didn't, parameters were adjusted. The underlying boundary between description and definition remained in place, but it was no longer examined. It became background structure.

The agreement remained silent because it did not need defending. As long as results kept arriving, no one had reason to question the terms. The discipline rewarded those who stayed within the boundary and quietly discouraged those who pressed against it. Not through censorship, but through indifference. Work that asked "what is really happening?" without offering immediate calculational payoff simply struggled to find traction.

This posture also shaped how Einstein himself was remembered. His restraint was reinterpreted as finality. His caution was mistaken for closure. Later generations spoke as if he had answered questions he had carefully avoided. Geometry was no longer just a language for consequences; it became the assumed fabric of reality. Expansion was no longer an interpretation of redshift; it became a literal description of the universe's behavior.

None of this required an explicit decision. It emerged naturally from the silent agreement: if a framework predicts accurately, it is treated as reality. If it does not, it is revised until it does. The question of whether the framework describes behavior or substance becomes secondary—often invisible.

This invisibility is what makes the agreement powerful and dangerous at the same time. Powerful, because it enables progress without philosophical gridlock. Dangerous, because it discourages reflection on assumptions that quietly accumulate. When interpretation hardens into belief, anomalies no longer invite rethinking; they demand correction.

Modern cosmology illustrates this tension clearly. Observational discrepancies are met with additional components rather than with interpretive pause. Dark Matter and Dark Energy are introduced not because they are directly observed, but because the existing framework requires them. Expansion rates are reconciled through increasingly subtle mechanisms, while the possibility that expansion itself is an interpretation remains largely untouched.

The silent agreement does not forbid questioning these moves. It simply makes such questioning feel unnecessary, even irresponsible. Why reopen foundations when the equations still work?

Yet this is precisely the moment when foundations matter most. When a framework requires ever more elaborate adjustments to preserve descriptive success, it signals not failure, but saturation. The language of consequences has reached its limit of flexibility. Continuing to stretch it without examining its assumptions risks turning discipline into dogma.

Einstein sensed this risk, even if he did not resolve it. His lifelong unease was not with calculation, but with interpretation. He accepted the silent agreement for what it was—a working truce, not a permanent settlement. His refusal to define the universe was not an evasion; it was a recognition that physics had not yet earned the right to do so.

The community, however, gradually forgot that the agreement was provisional. What began as methodological humility became an unspoken rule. As long as predictions held, questions about underlying reality were deferred indefinitely.

This book does not call for abandoning that agreement. It calls for remembering that it exists. Silence can be useful, but only if it is chosen consciously. When silence becomes habit, it stops protecting inquiry and starts limiting it.

Recognizing the silent agreement science made allows us to see both the strength and the fragility of modern physics. It explains how we achieved such precision without ontological clarity—and why, now, that clarity can no longer be postponed without cost.

Einstein drew a boundary and respected it. The community honored that boundary by not crossing it, but also by not naming it. The next step is not to rush past it, but to finally acknowledge it—and ask whether the time has come to move beyond it with the same restraint that created it in the first place.

1.5 — The Question Left Open

When all the restraint is accounted for, when the discipline is understood, and when the silence is finally acknowledged, a single question remains standing. It is not dramatic, and it is not new. It does not arise from dissatisfaction with results, but from their very success. It is the question Einstein left open—not because he overlooked it, but because he knew it could not yet be answered responsibly.

The question is simple to state and difficult to face: *what is it that responds?*

Einstein showed us how clocks slow, how paths bend, how light shifts, how motion reorganizes itself in the presence of energy. He mapped these responses with extraordinary precision. But in every case, something is doing the responding. Something accommodates distortion. Something allows redistribution. Something makes relational behavior possible at all. Einstein never named that "something."

This omission was not an accident. It was a recognition of limits. To name the responder would have required a physical claim about the nature of the vacuum, the substrate of space, or the medium of interaction—claims for which there was, at the time, no experimental grounding. Einstein understood that once such a claim is made, it shapes all subsequent interpretation. A wrong name is harder to undo than an unanswered question.

So he stopped where certainty ended.

Yet stopping does not mean erasing the question. It means postponing it. And postponed questions have a way of returning when the frameworks built around them begin to strain.

For decades, the question could be ignored because the language of consequences was sufficient. Predictions matched observation. New effects were explained without revisiting first principles. But as physics expanded its reach—from local systems to cosmic scales—the unanswered question quietly reasserted itself. The vacuum, once treated as nothing, was asked to do more and more. It was expected to curve, stretch, accelerate, and store energy, all without being defined.

At that point, silence ceased to be neutral.

The emergence of concepts like Dark Matter and Dark Energy did not signal discovery in the traditional sense. They signaled necessity. Something unseen was required to preserve the descriptive framework. Parameters were introduced not because the responder was identified, but because the response could no longer be accounted for otherwise. The question Einstein left open was being answered indirectly—through placeholders rather than understanding.

This is why the question could not be avoided forever. A framework that describes behavior without naming the responder can function only as long as behavior remains within certain bounds. When anomalies multiply, when constants refuse to remain constant, and when interpretations require increasing adjustment, the cost of not naming becomes visible.

The unresolved question is not "why do equations fail?" They largely do not. The question is "what exactly do these equations describe?" Are they telling us about geometry acting on its own? About space behaving like a substance? About nothingness exerting influence? Or are they mapping the behavior of a physical medium that has never been explicitly acknowledged?

Einstein did not answer this because he could not do so without crossing his own methodological boundary. But he also did not deny that such a question existed. His later work, his discomfort with closure, and his refusal to accept completeness all point to an awareness that description alone was not the end of inquiry.

What has changed since Einstein's time is not the validity of his equations, but the context in which they operate. We now probe deeper, farther, and with greater precision. We observe tensions that are not errors, but signals. We see patterns that suggest structure beneath what was assumed to be empty. The question of the responder is no longer speculative—it is unavoidable.

This does not mean Einstein failed to answer the question. It means he succeeded in leaving it open without corrupting the discipline. He created a framework strong enough to survive without ontology, and honest enough not to pretend it had one.

The closing of this chapter does not attempt to answer the question either. It simply names it clearly, perhaps for the first time without hesitation. *What is it that responds when energy is present? What is it that redistributes, equilibrates, and gives rise to the consequences we observe?*

Once that question is named, it cannot be unseen. It becomes the thread that connects every chapter that follows. Einstein stopped at the boundary because crossing it prematurely would have weakened physics. The work ahead begins not by leaping across that boundary, but by standing at it long enough to understand why it was drawn—and why, now, it may finally be possible to step beyond it without losing the discipline that made everything else work.

Chapter 2 — What Einstein Actually Solved

"The important thing is not to stop questioning. Curiosity has its own reason for existence." — **Albert Einstein**

Clearing the Fog Around the Achievement

To understand what Einstein actually solved, we must first clear away what has gathered around his name.

Over the decades, Einstein's work has been stretched into meanings he never claimed and responsibilities he never accepted. He is often portrayed as the man who explained the universe itself—its origin, its structure, even its destiny. This image is flattering, dramatic, and deeply misleading. It turns a disciplined scientific achievement into a mythological one.

Einstein did not solve the universe. He solved a very specific and very difficult problem: *how physical systems behave in the presence of energy and motion when classical assumptions fail.*

That difference matters.

The temptation to inflate his contribution comes from its success. General Relativity worked so well that it began to feel complete. Predictions matched observations with precision. New phenomena were explained without contradiction. In such an environment, it is easy to assume that what works must also explain everything. But effectiveness is not the same as exhaustiveness.

Einstein's real achievement was narrower—and far more subtle. He identified a consistent way to describe how matter and energy influence relationships between events: how time is experienced, how paths are followed, how measurements depend on context. He did not explain what time is. He showed how time behaves. He did not explain what space is. He showed how spatial relations change. He did not identify a physical substance behind gravity. He replaced force with response.

This chapter is about restoring that precision.

To strip away mythology is not to diminish Einstein's work. It is to respect it. When we attribute to him answers he never gave, we lose sight of the discipline that made his work powerful. We also inherit problems that were never his to solve. The confusion between what Einstein calculated and what later generations assumed is one of the quiet sources of tension in modern physics.

Einstein solved behavior, not being. He solved response, not substance. He solved consistency, not ontology.

Once this is understood, much of the confusion surrounding modern cosmology begins to untangle. The questions we now struggle with—about expansion, vacuum, dark components, and cosmic tension—do not arise because Einstein failed. They arise because his solution was asked to carry more meaning than it was designed to hold.

This chapter begins by placing Einstein's work back into its original frame. Not as a theory of everything, but as a theory of *how things respond.* Only then can we appreciate both the strength of what he achieved and the necessity of asking what lies beyond it.

2.1 — The Problem of Gravity Before Einstein

Before Einstein, gravity was not a mystery because it failed to work. It was a mystery because it worked *too well* without explaining itself. The problem was not observational, experimental, or mathematical. It was conceptual. Gravity behaved with relentless consistency while offering no clue about how it did so.

To understand why Einstein's intervention mattered, we must set aside historical storytelling and focus on the core difficulty gravity posed. The issue was never that objects fell or planets moved unpredictably. On the contrary, they did so with astonishing regularity. The deeper problem was that gravity acted without contact, without delay, and without a visible intermediary. It produced effects everywhere while appearing to originate nowhere in particular.

Classical physics tolerated this unease for a long time. Newton's formulation described gravitational behavior with unmatched precision. Objects accelerated toward one another according to mass and distance, and the equations held across terrestrial and celestial scales. Yet Newton himself was uncomfortable with what he had written. He famously refused to speculate on the cause of gravity. Action at a distance, instantaneous and unexplained, violated the mechanical intuitions of his era.

What troubled physicists was not the mathematics, but the metaphysics implied by it. How could one object influence another across empty space without anything in between? What carried the influence? What mediated the interaction? These were not idle questions. They pointed to a gap between description and understanding that physics had not yet bridged.

Various attempts were made to close this gap. Fields were introduced as conceptual tools—ways to represent influence without direct contact. Yet even fields, initially, were abstractions. They described regions of influence but did not necessarily correspond to physical substances. The ether was proposed as a medium to support these fields, but its experimental failure only deepened the confusion. Gravity, like light, seemed to require a carrier, yet none could be found.

By the late nineteenth century, gravity occupied an awkward position. It was indispensable, accurate, and conceptually unsatisfying. It did not fit neatly into the mechanical worldview that dominated physics. Other forces could be explained through interactions, collisions, or transmission. Gravity remained aloof, acting universally and silently.

The difficulty was not merely philosophical. It had practical consequences. Newtonian gravity implied instantaneous action, which conflicted with emerging ideas about finite signal speeds. It treated time and space as absolute backgrounds, unaffected by the very forces operating within them. These assumptions held well enough until experiments began probing regimes where precision exposed their limits.

The problem, then, was not that gravity lacked a formula. It lacked a *home*. It had no clear place within the conceptual architecture of physics. It acted on objects, yet seemed indifferent to the medium through which it acted. It depended on mass, yet did not arise from any known material process. It was everywhere and nowhere at once.

This is the problem Einstein confronted—not a failure of prediction, but a failure of coherence.

Einstein did not approach gravity by asking what force pulls objects together. He asked why objects follow the paths they do. This shift may sound subtle, but it changed everything. Instead of treating gravity as an interaction transmitted across space, he treated it as a condition of motion itself. Objects were not being pulled; they were responding to the structure of their environment.

By reframing the problem this way, Einstein sidestepped the need to explain how gravity travels. He replaced the question of transmission with a question of relation. Motion was no longer something imposed externally; it was something guided by context. Paths changed not because a force acted upon them, but because the conditions defining those paths had changed.

This move addressed the conceptual discomfort that had haunted gravity for centuries. If gravity was not a force in the traditional sense, then the demand for a carrier weakened. There was no need to imagine invisible threads pulling objects together. Instead, objects moved as they did because the rules governing motion were different in the presence of mass and energy.

What Einstein solved, at this stage, was not the nature of gravity, but its inconsistency with the rest of physics. He aligned gravity with the principle that nothing propagates instantaneously. He reconciled it with the relativity of time and space. He removed action at a distance by removing the need for action altogether.

Yet this solution came with a price—or rather, with a limit.

By relocating gravity into the geometry of relations, Einstein avoided defining what, if anything, underlies those relations. He provided a consistent way to calculate motion without naming a physical medium. Gravity ceased to be mysterious because it ceased to be an interaction. But the question of *what responds* remained unanswered.

This was not a failure. It was a disciplined choice. The pre-Einstein problem of gravity was conceptual confusion masquerading as physical mystery. Einstein dissolved that confusion by changing the question. He solved the problem gravity posed *to physics*, not the problem gravity posed *to ontology*.

Understanding this distinction is essential. Einstein did not answer what gravity is. He answered how gravitational behavior fits coherently within a relativistic universe. He gave gravity a place in the language of physics, even if he did not give it a substance.

The difficulty gravity posed before Einstein was never about falling apples or orbiting planets. It was about how something could influence everything without being anything in particular. Einstein resolved that difficulty by refusing to speculate beyond what could be described. He gave physics a way to move forward without pretending that the deeper question had been settled.

That deeper question—what, if anything, underlies gravitational response—was left intact. It would resurface later, not as a problem with Einstein's solution, but as a consequence of its success.

This chapter does not judge that choice. It recognizes it as the only choice available at the time. To understand what Einstein actually solved, we must first understand what he deliberately did not attempt to solve. Gravity before Einstein was a conceptual orphan. Einstein gave it coherence without claiming parentage.

2.2 — From Force to Response

Einstein's most radical move was not mathematical. It was conceptual. He did not begin by asking how gravity acts; he began by questioning whether it should be treated as an *action* at all. This shift—from force to response—quietly dissolved problems that had resisted centuries of refinement.

In classical physics, force is an external agent. It acts on objects, alters their motion, and does so through some form of interaction.

Even when the mechanism is unknown, the structure of explanation remains the same: something pushes, pulls, or transmits influence. Gravity, under this view, was an interaction without a visible intermediary, an exception tolerated only because it worked.

Einstein did not try to improve that interaction. He abandoned the framework that required it.

Instead of asking what pulls objects together, he asked why objects move the way they do when energy and motion are present. This may appear like a change in wording, but it was a change in explanatory posture. Motion was no longer something imposed from outside; it became something that followed naturally from the conditions defining the system.

In this reframing, gravity ceased to be a force acting *on* objects and became a pattern of response *by* objects. Objects did not accelerate because they were pulled. They accelerated because the rules governing motion had changed. Paths that appeared straight in one context were no longer straight in another. What looked like force was reinterpreted as geometry—not geometry as substance, but geometry as relational structure.

This is where the language of response becomes crucial. A response does not require a messenger. It does not demand a carrier. It does not travel from source to target. A response occurs because conditions exist. When temperature rises, materials expand. When pressure changes, fluids flow. No one asks what "transmits" expansion or flow. They are responses to altered conditions.

Einstein placed gravity into this category.

Mass and energy altered the conditions under which motion occurred. Objects responded by following different trajectories. Time responded by flowing differently. Distances responded by being measured differently. Nothing was pushed. Nothing was pulled. The universe simply behaved differently under different conditions.

This reframing solved multiple conceptual problems at once. It removed the need for instantaneous action at a distance. It aligned gravity with the principle that nothing propagates faster than light, not by slowing gravity down, but by eliminating propagation altogether. Gravity was no longer something that traveled; it was something that was already there, encoded in the structure governing motion.

What Einstein replaced was not force with geometry, but interaction with context. Geometry was the mathematical language capable of expressing that context with precision. It allowed the rules of motion to vary smoothly across space and time without introducing new agents. In doing so, it transformed gravity from an awkward exception into a natural consequence of relational physics.

This is why Einstein's solution feels inevitable in hindsight. Once motion is understood as dependent on context, force becomes optional. Acceleration no longer demands an external cause; it can arise from the way paths are defined. The familiar sensation of being pulled downward becomes a local experience of responding to conditions, not evidence of an invisible hand.

Yet this elegance came with an implicit limitation.

By redefining gravity as response, Einstein deliberately avoided naming what establishes the conditions to which objects respond. Geometry described how those conditions vary, not what gives rise to them. The equations told us how motion unfolds, not what underwrites the structure within which it unfolds.

This restraint was consistent. Einstein did not claim that spacetime *acts*. He showed that spacetime *describes* how motion responds. Later language often blurred this distinction, speaking as if spacetime bends objects or tells matter how to move. Einstein himself was careful. The response was primary; agency was not assigned.

Understanding this difference clarifies what Einstein actually achieved. He did not demystify gravity by explaining its cause. He made gravity intelligible by removing the need for a cause in the traditional sense. He replaced the question "what force is acting?" with "what conditions define motion here?"

That move was both powerful and incomplete—by design.

It was powerful because it aligned gravity with the broader shift in physics away from absolute structures and toward relational descriptions. Motion became contextual, not absolute. Time became local, not universal. Space became dynamic, not fixed. Gravity fit naturally into this framework once it was treated as response.

It was incomplete because response, by definition, presupposes something that responds. Einstein described the pattern of response with extraordinary clarity, but he did not identify the responder. Geometry encoded the rules, but it did not explain why those rules exist or what enforces them physically.

Einstein accepted this incompleteness consciously. He chose coherence over closure. He allowed gravity to be understood operationally without pretending that its physical substrate had been identified. This choice protected his theory from speculative excess and made it robust across domains.

Later generations often forgot that this was a choice. Response was reinterpreted as action. Geometry was treated as substance.

Spacetime was spoken of as if it possessed agency. These shifts were not present in Einstein's original framing; they emerged as interpretive shortcuts.

Returning to the original move—from force to response—restores the precision of Einstein's achievement. He did not weaken gravity by removing force; he strengthened it by making it consistent with the rest of physics. He showed that what we experience as gravitational attraction is not an interaction demanding explanation, but a response demanding description.

This section matters because it sets the boundary for everything that follows. Einstein solved the problem of gravity by changing how we think about motion itself. He did not solve the problem of what establishes the conditions of response. That question was not ignored. It was postponed.

Understanding that postponement is essential. Without it, we risk attributing to Einstein answers he never claimed to give—or blaming him for questions he deliberately left open.

2.3 — Field Behavior Around Mass

When stripped of symbolism, mythology, and later interpretation, Einstein's field equations do something remarkably specific. They do not describe what the universe is made of. They do not introduce a new substance or agent. They answer a narrower but decisive question: *given a concentration of energy, how does the surrounding environment behave?*

This focus is easy to overlook because the equations are often presented as statements about spacetime itself. Yet their operational role is not to define spacetime as an entity, but to relate energy distribution to observable effects in its neighborhood. The equations are, at their core, behavioral rules.

What they encode is not gravity as a force, but *how conditions change around mass and energy*, and how those changes affect motion, timing, and measurement.

Consider what happens when mass is present. Clocks tick differently nearby. Paths that would otherwise appear straight begin to curve. Light changes direction. Orbits precess. None of these effects require objects to be pulled. They require only that the conditions governing motion are no longer uniform. The field equations specify how those conditions vary.

In practical terms, the equations function like an instruction set for the environment. Given a certain energy–momentum distribution, they tell us how intervals, trajectories, and rates adjust locally. They translate energy concentration into altered relational behavior. That translation is their achievement.

This is why the equations are indifferent to interpretation. Whether one speaks of curvature, distortion, or geometry is secondary. What matters is that the equations reliably predict *responses*. They tell us how matter and radiation behave in the vicinity of mass, not because spacetime exerts a force, but because the rules that define motion and measurement have shifted.

This becomes clearer when we look at how the equations are used. In application, physicists do not ask what spacetime is doing. They ask what will happen to clocks, signals, and trajectories near a given mass. The equations are solved to obtain observable outcomes: time dilation factors, deflection angles, orbital corrections. These are all consequences, not substances.

Even the term "field" can mislead. In Einstein's usage, the gravitational field is not a physical medium in the classical sense. It is a mathematical representation of how relations vary across regions. It encodes influence without specifying a carrier. The field exists as a rule-set, not as a material presence.

This is why General Relativity fits comfortably alongside other consequence-based frameworks. Maxwell's equations predict how electric and magnetic effects manifest around charges. They do not require a physical ether to function. Similarly, Einstein's equations predict how gravitational effects manifest around energy concentrations without requiring a physical spacetime substance. In both cases, the equations govern behavior, not ontology.

The predictive power of this approach is what made it irresistible. Once physicists realized that they could calculate observable outcomes with unprecedented accuracy, the temptation to ask deeper questions diminished. Why inquire into what underlies the field if the field equations already deliver correct answers?

Einstein himself never mistook this success for completeness. He was careful to emphasize that the equations relate geometry to energy, not geometry to substance. They connect measurable quantities—energy, momentum, curvature—without claiming that curvature itself is a thing. The equations do not say "mass creates spacetime." They say, more modestly, that energy distribution and relational structure are linked.

Understanding this distinction resolves a great deal of confusion. When people say that mass "warps spacetime," they often imagine a physical deformation, as though space were a fabric. This imagery is pedagogical, not literal. What the equations actually provide is a mapping from energy to altered relational behavior. They tell us how measurements change, not what bends.

This is also why the equations remain valid even as interpretations evolve. One can reinterpret what curvature represents without changing the equations themselves, because the equations were never committed to a particular ontology. They operate at the level of consequence.

But this strength also reveals the limit of what was solved.

The equations predict how surroundings behave near mass. They do not explain *why* energy concentration produces these effects. They specify the relationship, not the mechanism. They tell us that when energy is present, the environment responds in a particular way—but they do not identify what that environment physically is.

This is not an oversight. It is the logical outcome of Einstein's approach. Having reframed gravity as response rather than interaction, he no longer needed to specify a medium. The equations functioned perfectly without one. To name the medium would have been to cross the boundary he had carefully respected.

Yet the absence of a named responder remains noticeable once one looks closely. The equations imply responsiveness without identifying the responder. They describe behavior without grounding it in a physical substrate. As long as the predictions held, this gap was acceptable. As physics pushed into regimes where interpretation mattered more—cosmology, vacuum energy, large-scale structure—the gap became harder to ignore.

Recognizing what the field equations actually do is essential to understanding both their success and their limitations. They are not metaphysical statements. They are behavioral prescriptions. They tell us how the universe reacts locally to energy concentration, not what the universe fundamentally consists of.

Einstein solved the problem of predicting gravitational behavior by encoding it into a relational framework. He did not solve the problem of what carries or enables that behavior. That question lies beyond the equations by design.

This section closes on that recognition. Einstein's equations give us the most accurate map of gravitational response ever constructed. They were never meant to tell us what the terrain itself is made of.

2.4 — Why This Was Enough

From a scientific standpoint, Einstein did not need to go further than behavior. Solving how systems respond under given conditions was not a partial victory—it was exactly what experimental science required. Anything beyond that would have added philosophical weight without increasing predictive power. In that sense, Einstein stopped not because he fell short, but because he had reached the threshold where science, as it was practiced, could advance no further without changing its nature.

Experimental success depends on one thing above all else: reliable prediction. A theory earns its place not by explaining reality in absolute terms, but by telling us what will happen when we perform an experiment. Will a clock tick slower? By how much? Will light bend near a massive body? At what angle? Will an orbit precess? By what degree? Einstein's framework answered these questions with extraordinary precision.

Once those answers were confirmed, the theory had done its job.

It is important to recognize how modest experimental demands actually are. Experiments do not test what something *is*. They test what something *does*. A detector does not care whether gravity is a force, a curvature, or a response of an unseen medium.

It records outcomes. If a model reproduces those outcomes consistently, it passes.

Einstein understood this deeply. He did not burden his theory with unnecessary claims because those claims would not improve its testability. Naming an underlying substance for spacetime would not have changed a single prediction. It would not have sharpened a measurement. It would not have altered an experimental result. From a strictly scientific perspective, it would have been excess.

This is why the absence of ontology did not weaken General Relativity—it strengthened it. By restricting the theory to relationships between measurable quantities, Einstein ensured that it could be tested cleanly. There was nothing extraneous to falsify. Either the predicted behavior occurred, or it did not. And time and again, it did.

The sufficiency of behavior-based explanation becomes even clearer when we consider how experiments are designed. No experiment ever isolates "spacetime itself." Experiments isolate clocks, rulers, particles, signals, and trajectories. What changes is not spacetime as an object, but the readings of instruments placed in different conditions. Einstein's equations were built precisely to predict those readings.

This alignment between theory and experiment is not accidental. It reflects a deliberate choice to keep the theory operational. Every term in the equations corresponds, directly or indirectly, to something that can be measured. Energy and momentum describe physical content. Curvature encodes relational change. The link between them yields observable effects. Nothing in this chain requires naming a substance.

As long as this chain remained intact, the theory could not fail experimentally. And for a long time, it did not.

Moreover, behavior-based success has a stabilizing effect on science as a community practice. When a theory works across multiple independent tests, confidence builds. Resources flow toward refinement rather than reconsideration. Interpretive questions recede because they are not required for progress. The discipline moves forward by extending application rather than revisiting foundations.

Einstein's solution fit perfectly into this mode of progress. It solved a pressing conceptual inconsistency in gravity while delivering precise, testable predictions. It did so without introducing speculative entities or mechanisms that would have complicated verification. From the standpoint of scientific economy, it was ideal.

This is why later generations felt no urgency to ask what lay beneath the equations. The framework was productive. It unified disparate phenomena. It integrated gravity with relativistic motion. It did not demand completion in order to function. The success itself became justification for silence.

Even cosmology initially benefited from this restraint. Large-scale models could be built by extending the same relational logic outward. Redshift could be interpreted within the same framework. Cosmic dynamics could be described without redefining the underlying medium. As long as observations could be matched, the approach held.

What matters here is not that Einstein's theory lacked something, but that it lacked nothing *necessary* for the tasks it was asked to perform. Experimental science does not reward completeness; it rewards consistency. Einstein delivered consistency at a level unmatched before or since.

The reason this was enough, for so long, is that science had not yet reached the limits of descriptive elasticity. The consequences being modeled remained within a range that could be accommodated without naming deeper structure. Behavior alone could carry the explanatory load.

But sufficiency is not permanence.

A theory can be sufficient for one era and strained in another without ever being wrong. When new regimes are explored, when precision tightens, and when interpretation begins to matter as much as calculation, the question of underlying structure resurfaces. Not because behavior was misdescribed, but because behavior alone no longer satisfies understanding.

Einstein anticipated this moment, even if he did not resolve it. His refusal to declare finality was not philosophical hesitation; it was scientific foresight. He knew that a theory built entirely on consequences could succeed brilliantly and still leave something essential untouched.

Understanding why his solution was enough allows us to appreciate both its power and its limits. It explains why Einstein's work dominated physics for a century—and why, now, the same success invites a deeper question.

The next section turns to that boundary: how solving behavior with such precision created the illusion of completeness, and how easily sufficiency can be mistaken for final explanation.

2.5 — Precision Without Ontology

Einstein demonstrated something that is still difficult for many to accept: one can be exactly right about outcomes while remaining deliberately silent about causes. This is not a weakness of science. It is one of its most powerful disciplines.

Precision does not require ontology. It requires consistency.

Einstein's equations predict outcomes with extraordinary accuracy. They tell us how clocks diverge, how light bends, how orbits evolve, how signals delay. These predictions do not depend on knowing what spacetime *is*. They depend only on knowing how measurable quantities relate under given conditions. As long as those relations are correct, the outcomes will be correct—regardless of what lies beneath them.

This is why General Relativity could pass every experimental test thrown at it without ever naming a physical substrate. The theory never promised to explain causes. It promised to predict consequences. And it kept that promise.

This separation between outcome and cause is not accidental. It is the result of a disciplined refusal to speculate where speculation would not improve testability. A cause that cannot be independently observed or constrained adds nothing to prediction. It may satisfy curiosity, but it does not strengthen the theory. Einstein understood this distinction clearly.

In practice, experiments do not interrogate causes. They interrogate responses. A gravitational lens does not reveal what space is made of. It reveals how light behaves near mass. A clock on a satellite does not expose the nature of time. It shows how time is measured differently under motion and gravity. These are outcomes, and Einstein's framework was built to capture them precisely.

What makes this remarkable is that silence about causes did not reduce explanatory power at the level science required. On the contrary, it increased it. By refusing to anchor his theory to an unverified ontology, Einstein ensured that it remained flexible, resilient, and widely applicable. The theory could be extended, reinterpreted, and adapted without breaking its core predictions.

This is why later discoveries did not invalidate Einstein's work. New observations refined parameters, not principles. Even when General Relativity was embedded within larger cosmological models, its predictive core remained untouched. Precision endured because it was never tied to a fragile definition of reality.

However, precision without ontology carries a psychological cost. When predictions work repeatedly, the silence about causes begins to feel unnecessary. The absence of explanation is mistaken for explanation itself. The language used to describe outcomes is quietly promoted to the status of reality. Curvature becomes substance. Expansion becomes action. Geometry becomes actor.

This shift is subtle, and it is understandable. Humans prefer stories with agents and mechanisms. A theory that predicts perfectly but explains modestly feels incomplete. Over time, that discomfort seeks resolution—not by revisiting foundations, but by reinterpreting descriptions as causes.

Einstein resisted this temptation throughout his life. He never claimed that spacetime itself was a physical thing exerting influence. He did not say that geometry *causes* motion. He said that geometry *describes* how motion unfolds in the presence of energy. The distinction is small in language and enormous in meaning.

By maintaining this restraint, Einstein preserved the integrity of his theory. He allowed it to be precise without being presumptuous. He showed that physics could achieve unparalleled accuracy without declaring metaphysical victory.

The cost of ignoring this lesson becomes visible when precision is mistaken for completeness. When outcomes are treated as explanations, anomalies are handled by adjustment rather than inquiry. When silence about causes is forgotten, the theory is burdened with interpretations it was never designed to support.

Understanding precision without ontology restores balance. It allows us to respect Einstein's achievement without inflating it. It explains why his work remains valid even as questions accumulate around it. It clarifies how a theory can be exactly right about what happens while remaining silent about what makes it happen.

This silence was not an evasion. It was a choice.

Einstein solved what could be solved with the tools available. He refused to name what could not yet be grounded physically. His precision was complete where it mattered—at the level of consequence. His ontology remained open where honesty demanded it.

This closing recognition prepares the ground for what follows. If one can be precisely right about outcomes without naming causes, then the task ahead is not to correct Einstein, but to ask whether the time has come to name what he left unnamed—without sacrificing the discipline that made his precision possible.

Chapter 2 ends here, not with a conclusion about reality, but with a clarified understanding of achievement. Einstein solved behavior completely. He left being deliberately undefined. The chapters ahead begin where that silence becomes impossible to ignore.

Interlude I — What Does It Mean to Explain?

Before moving forward, it is worth pausing—because much of what follows depends on a quiet distinction that often goes unnoticed. We frequently use the words *explain* and *describe* as if they were interchangeable. In practice, they are not. Understanding the difference between them changes how we read science, how we interpret evidence, and how we evaluate truth itself.

To describe something is to account for how it behaves. To explain it is to reveal why that behavior occurs. The two often travel together, but they are not the same journey.

A description tells us *what happens*. It captures patterns, regularities, and relationships. It allows prediction. It offers control. Explanation, on the other hand, seeks grounding. It asks what gives rise to those patterns in the first place. Description organizes phenomena; explanation tries to locate their origin.

Modern science excels at description. Its mathematical frameworks are extraordinarily effective at capturing how systems behave under a wide range of conditions. From planetary motion to quantum transitions, we can predict outcomes with astonishing accuracy. Yet this success can blur an important distinction: accuracy does not automatically imply understanding.

When a model predicts reliably, we tend to treat it as explanatory—even if it does not address the deeper question of *why* the model works. Over time, the model's success grants it authority, and its assumptions become invisible. Description hardens into explanation without ever being examined as such.

This shift is subtle, and it happens not through error but through triumph. When a framework works consistently, questioning its foundations can feel unnecessary, even disruptive. The language of the model becomes the language of reality. Concepts that began as tools become treated as truths.

This is where reflection becomes essential.

Explanation, in its deeper sense, does not merely connect variables. It reveals the conditions under which those variables make sense. It clarifies what must be true for a given description to hold. Without this step, understanding remains functional but incomplete—powerful, yet ungrounded.

Science, at its best, oscillates between these two modes. It builds descriptions, then steps back to examine what those descriptions imply. Progress occurs not only through accumulation of data, but through moments when we realize that our most successful explanations are still approximations.

This interlude invites that pause.

It does not challenge the validity of established theories. It questions the level at which they operate. It asks whether we have mistaken effective description for foundational understanding, and whether the line between the two has quietly blurred over time.

Such reflection is not a retreat from rigor; it is a return to it. The history of science shows that the deepest advances often come not from new measurements, but from new ways of interpreting what we already know.

As we move forward, the aim is not to discard existing explanations, but to see what they leave unsaid. To ask what lies beneath the language we have grown accustomed to using. To recognize that explanation is not a destination, but a process—one that evolves as our questions deepen.

This pause, then, is not a detour. It is a recalibration. A moment to realign our understanding before continuing the journey with clearer eyes and sharper intent.

Part II — The Language Of Description

How modern physics came to speak the way it does.

Chapter 3 — Geometry as a Language of Consequences

"Concepts which have proved useful in ordering things easily achieve such an authority over us that we forget their earthly origins and accept them as unalterable givens." — **Albert Einstein**

When the Language Started Sounding Like Reality

At some point, geometry stopped sounding like a language and began to sound like a thing.

What was introduced as a way to describe relationships gradually acquired the weight of substance. Curvature was no longer just a mathematical expression of how motion behaves; it became something that was said to exist, to act, and to govern. Spacetime, once a framework for measurement, began to be spoken of as if it were a physical fabric capable of bending, stretching, and evolving on its own.

Einstein never insisted on this shift.

He introduced geometry because it was the most precise grammar available for expressing relational change. It allowed him to encode how measurements depend on context without inventing new forces or entities. Geometry, in his hands, was not a claim about what reality is made of. It was a way of writing down how reality behaves.

Yet success has a way of hardening tools into truths. As predictions accumulated and experiments confirmed them, the distinction between description and substance blurred. The language worked so well that it began to feel sufficient as explanation. Geometry, which had started as syntax, quietly assumed the role of actor.

This chapter returns geometry to its original place.

Not to weaken its importance, but to clarify its function. Geometry did not replace physics; it expressed it. It did not introduce a new substance; it organized consequences. Understanding this difference is essential, because much of modern confusion arises not from faulty equations, but from mistaking the grammar of a theory for the reality it describes.

Before we ask what space and time are, we must first understand how and why geometry became the language through which their behavior was written—and how easily a language, once successful, can be mistaken for the thing it names.

3.1 — Why Geometry Was Chosen

Geometry was not chosen because it was fashionable, abstract, or elegant. It was chosen because it was unavoidable.

By the time Einstein began rethinking gravity, physics had reached a point where traditional language was no longer sufficient. Concepts like force, trajectory, and absolute motion had served well in a stable, background universe. But once motion itself became relative—once time and distance depended on the observer—those concepts began to lose their clarity. Physics needed a language that could express relationships without assuming fixed backgrounds.

Geometry was already that language.

For centuries, geometry had been used to describe relationships independent of material substance. It did not ask what things were made of; it described how they were arranged. Distance, angle, curvature, and continuity were relational concepts by nature. They did not require objects to act. They described how positions and paths relate under given rules. This made geometry uniquely suited to a physics that was shifting away from absolute structures.

Einstein's insight was to recognize that gravity was not resisting explanation because it was mysterious, but because it was being asked the wrong kind of question. Treating gravity as a force required identifying an agent and a mechanism. Treating gravity as a modification of relational structure required only a way to describe how relations change. Geometry provided exactly that.

Once motion was understood as contextual rather than absolute, geometry ceased to be optional. If time flows differently depending on conditions, and if paths change not because something pushes but because the rules of motion vary, then those rules must be expressed relationally. Geometry is the discipline that does precisely this. It encodes how measurements depend on position, motion, and energy without invoking material causation.

This is why Einstein did not invent a new physical substance to explain gravity. He did not need to. The problem was not that gravity lacked a carrier; it was that the framework describing motion assumed a fixed stage. Geometry allowed the stage itself to be described as part of the play—without turning it into an actor.

Another reason geometry was chosen lies in its neutrality. Geometry does not claim agency. A curved surface does not cause motion; it defines paths. A shortest path on a curved surface is not imposed by force; it follows from structure. This analogy was powerful because it reframed gravitational motion as natural rather than compelled. Objects moved as they did because that was how motion unfolded under given conditions.

Einstein saw that this approach aligned perfectly with his broader view of physics. He was deeply uncomfortable with explanations that relied on unseen agents acting behind the scenes. Geometry allowed him to describe outcomes without inventing causes. It was descriptive without being speculative.

This descriptive power mattered enormously. Geometry could incorporate changes smoothly. It could handle variation across space and time without introducing discontinuities or exceptions. Where force-based explanations struggled with universality, geometry thrived. It did not matter whether one was near Earth, a star, or an empty region of space. The same relational rules applied everywhere, adjusted only by local conditions.

Equally important was geometry's compatibility with measurement. Geometry speaks the language of rulers and clocks. It deals directly with intervals, durations, and paths—the very quantities experiments actually measure. Einstein's theory was built to predict how these measurements change, not to speculate about what lies beneath them. Geometry was the most direct way to connect theory to observation.

It is also crucial to recognize what geometry was *not* chosen for. It was not chosen because space was thought to be a substance. It was not chosen to replace matter with mathematics. It was not chosen to declare that reality is geometric. These interpretations came later, as metaphors hardened into beliefs.

Einstein himself was cautious about such readings. He treated geometry as a means of expression, not as an ontological claim. For him, geometry was a way to encode the behavior of physical systems under relativistic conditions. It described how measurements relate, not what existence consists of.

The success of this choice cannot be overstated. Geometry allowed gravity to be integrated seamlessly into a relativistic framework. It eliminated the need for special cases. It preserved consistency across scales. Most importantly, it allowed physics to progress without answering questions it was not yet equipped to answer.

But success carries risk.

Because geometry worked so well, it began to feel fundamental. The distinction between language and reality blurred. Curvature, once a descriptor, began to be spoken of as if it were a physical deformation. Spacetime, once a framework, began to sound like an entity. These shifts were not part of Einstein's original intent; they were the byproducts of prolonged success.

Understanding why geometry was chosen helps prevent this confusion. Geometry was selected not because reality is geometric, but because geometry is the most efficient way to describe relational change without committing to speculative causes. It was chosen because it allowed physics to stay honest.

This section restores that honesty. Geometry was a tool—an extraordinarily powerful one—but still a tool. It was the grammar in which consequences were written, not the substance from which reality was built.

Once this is understood, geometry regains its proper place: indispensable, precise, and descriptive—yet silent about what ultimately lies beneath the relationships it so elegantly encodes.

3.2 — Curvature as Expression, Not Entity

Curvature, in Einstein's work, was never meant to be an object one could point to. It was a way of speaking about change—specifically, about how relationships between measurements vary from place to place. Yet over time, curvature has come to be treated as if it were a physical entity, something that exists independently and acts upon matter. This shift has been subtle, but it has had far-reaching consequences.

To understand what curvature means operationally, we must return to how it is actually used. Curvature does not appear in experiments as a thing. It appears only through effects. A clock ticks more slowly near a massive body. A light ray deviates from a straight path. An orbit shifts slightly from its classical prediction. These are not measurements of curvature itself. They are measurements of behavior. Curvature is the name we give to the pattern that makes these behaviors consistent.

In this sense, curvature is not observed; it is inferred. It is a compact way of summarizing how measurements relate to one another across a region. When those relations change smoothly, we describe the situation as curved. When they do not, we describe it as flat. Nothing material needs to be bent for this description to hold. What bends are expectations about how measurements would compare if conditions were uniform.

An operational understanding of curvature avoids metaphysical commitments. It does not ask what space is doing. It asks how paths, durations, and intervals compare when energy is present. If the comparison deviates from what would be expected in a uniform setting, curvature is the language used to encode that deviation.

This is why the familiar images of rubber sheets and depressions are misleading. They suggest that space is a substance that deforms under weight, like fabric under a load. These analogies are pedagogical conveniences, not physical claims. They help visualize relational change, but they quietly smuggle in an ontology that Einstein never endorsed.

In practice, curvature functions like a rulebook. It tells us how to translate local measurements into global expectations. It dictates how straight lines are defined in a given context, how clocks synchronize, how signals propagate. It does not push or pull. It constrains.

This distinction is crucial. A force acts. A rule constrains. Curvature belongs to the second category. It shapes behavior without exerting agency. Objects follow curved paths not because something bends them, but because those paths are the natural ones under the prevailing relational structure.

Consider a simple analogy that avoids physical metaphors. In a system of accounting, rules determine how transactions are recorded. If the rules change, the same transaction yields different balances. No one imagines that the rules act on the money. They define how outcomes are computed. Curvature plays a similar role. It defines how outcomes are computed for motion and measurement.

This is why Einstein could reformulate gravity without introducing a force. By expressing gravitational effects as curvature, he replaced an acting agent with a descriptive framework. Motion remained lawful, but the law was embedded in structure rather than imposed externally.

The confusion arises when curvature is reified—when it is treated as something that exists independently and causes effects.

This reification is tempting because curvature appears on the left side of equations, while energy appears on the right. The visual symmetry suggests a causal symmetry. But mathematically paired quantities are not necessarily ontologically equivalent. The equations relate descriptors, not substances.

Operationally, curvature is a bookkeeping device for relational change. It encodes how the presence of energy alters the comparison between measurements made at different locations or times. It does not require that space be a physical medium capable of stress or strain. It requires only that measurements are context-dependent.

Einstein was careful on this point. He did not say that matter bends space as an object bends a surface. He said that the presence of energy is associated with a change in the geometric description of relations. The difference is subtle in language and profound in meaning.

Treating curvature as expression rather than entity also explains why the theory remains valid across interpretations. One can adopt different views about what underlies curvature—vacuum structure, field dynamics, emergent behavior—without altering the predictive core of the equations. The equations care only that the relational description matches observation.

This flexibility is not a weakness. It is a sign that curvature was never meant to carry ontological weight. It was designed to be a neutral language that could survive reinterpretation. When later generations treat curvature as substance, they narrow that flexibility and inherit problems the original formulation avoided.

Operational clarity dissolves many apparent paradoxes. When asked how space can expand, curve, or ripple, the answer becomes simple: space does none of these things as an object. What changes are the relations by which we compare distances, durations, and paths. Curvature is the name we give to that change.

Seen this way, curvature is not mysterious. It is economical. It compresses a complex pattern of relational variation into a single descriptive term. It allows calculation without metaphysics. It permits precision without commitment.

This is why curvature should be understood as expression, not entity. It expresses how the environment behaves near energy concentrations. It does not exist as a thing alongside matter and radiation. It is a language for consequences, not a component of reality.

Restoring this operational meaning does not weaken Einstein's work. It clarifies it. It returns geometry to its proper role—as grammar, not substance. And once geometry is understood this way, the path opens to ask a deeper question without confusion: if curvature expresses response, what is it that responds?

That question lies beyond geometry itself. Geometry can express response perfectly while remaining silent about the responder. Einstein accepted that silence. Understanding curvature as expression allows us to do the same—until we are ready to ask what, if anything, gives rise to the structure geometry so elegantly describes.

3.3 — When Language Becomes Reality

A descriptive language does not announce when it begins to replace what it describes. The transition is gradual, almost polite. It happens not through argument, but through familiarity. When a tool works consistently, it stops feeling like a tool. It starts feeling like the thing itself.

This is how language becomes reality in science.

Einstein introduced geometry as a way to write down relationships. Curvature, intervals, and paths were not meant to be entities; they were symbols that allowed consequences to be expressed coherently. For a long time, this distinction held. Geometry was understood as a formal structure, not a physical claim. But success has a quiet persuasive power. When predictions continue to work, year after year, across experiments and scales, the language that produced them gains authority.

Eventually, authority becomes assumption.

The shift did not occur because anyone explicitly decided that geometry *is* reality. It occurred because there was no immediate reason to resist the idea. When one says that light bends because spacetime is curved, the sentence sounds explanatory. It feels complete. The grammar offers closure, even if the substance does not. Over time, such sentences are repeated, refined, and simplified. Metaphors harden. What began as shorthand becomes belief.

This process is not unique to relativity. It has occurred throughout the history of physics. Forces were once treated as literal pushes. Fields were later treated as physical entities rather than mathematical constructs. Probabilities in quantum mechanics are often spoken of as if they exist independently of measurement. In each case, language that was introduced to manage complexity gradually acquired ontological weight.

What makes the case of spacetime geometry distinctive is its subtlety. Geometry is abstract by nature. It does not look like matter. It does not behave like an object. Precisely because it lacks obvious substance, it can be imagined to underlie everything without resistance. When geometry becomes reality, it does so quietly, without the friction that accompanies more tangible claims.

The educational process reinforced this transition. Diagrams, analogies, and simplified explanations were designed to make relativity accessible. Rubber-sheet models, bending grids, and visual metaphors helped students grasp relational change. But metaphors, once internalized, are difficult to unlearn. Over time, the teaching aids began to shape intuition more strongly than the original discipline they were meant to serve.

The distinction between "this is how we describe behavior" and "this is what exists" faded in everyday scientific language. Papers spoke of spacetime dynamics. Popular accounts described space stretching and time flowing differently. Even technical discussions adopted shorthand that implied agency. Geometry, which was never meant to act, was now described as if it did.

This is not a failure of intelligence. It is a feature of human cognition. We understand the world by attributing structure, and structure easily becomes substance in the mind. When a description is consistent, we treat it as real. When it is elegant, we trust it. When it predicts correctly, we stop questioning its scope.

Einstein himself did not make this leap. He remained careful in how he spoke about his own work. He resisted reifying geometry. He did not describe spacetime as a material fabric. He emphasized relationships, not entities. His caution, however, did not scale with his fame. The community adopted his equations more quickly than his restraint.

As the decades passed, the language of geometry began to do explanatory work it was never designed to do. Questions about cause were answered with geometric description. Why does light bend? Because spacetime is curved. Why do clocks slow? Because time itself is affected. Why does the universe expand? Because space is expanding. Each answer sounds self-contained, yet each merely restates the description in different words.

This is the point at which language stops guiding understanding and starts replacing it.

Once geometry is treated as reality, certain questions become meaningless by definition. Asking what underlies curvature sounds misguided if curvature itself is taken as fundamental. Asking what space is made of seems unnecessary if space is assumed to be nothing more than geometry. The descriptive framework becomes closed, not because it is complete, but because it has been promoted to ontology.

The consequences of this promotion are subtle at first. They appear as interpretive habits rather than explicit claims. But over time, they shape how anomalies are handled. When observations do not fit neatly, the geometry is adjusted. New components are added. Additional curvature is invoked. The language is stretched rather than questioned.

This stretching works—up to a point. But each adjustment reinforces the assumption that geometry itself is the thing doing the work. The deeper question of what makes geometry responsive is postponed indefinitely.

Tracing this transition matters because it explains why modern physics can feel both extraordinarily precise and conceptually unsettled at the same time. The equations are not failing. The language is being asked to do more than it was meant to do. Descriptive tools have hardened into assumed physical truths, and that hardening limits the kinds of questions that feel permissible.

Reversing this process does not require rejecting geometry. It requires remembering why it was chosen in the first place.

Geometry was selected because it could express relational change without committing to substance. When it is treated as substance, that original advantage is lost.

This section does not accuse the scientific community of error. It describes a natural evolution of understanding. Tools that work become trusted. Trusted tools become invisible. Invisible assumptions become reality. Recognizing this progression allows us to step back without dismantling what has been built.

Once language is returned to its proper role—as description rather than declaration—the door reopens to questions that were never answered, only deferred. Geometry can resume its role as grammar. Reality can be asked about again without confusion.

That reopening is essential for what follows.

3.4 — The Power and Danger of Mathematical Elegance

Mathematical elegance is one of science's most persuasive virtues. When equations are compact, symmetric, and internally consistent, they inspire confidence that extends beyond calculation. Elegance suggests depth. It hints at inevitability. It creates the impression that a theory is not merely useful, but true in some fundamental sense. This impression is powerful—and it is precisely why elegance can become dangerous.

Einstein's equations are elegant in the deepest way mathematics allows. They compress complex behavior into a minimal form. They unify diverse phenomena without special cases. They replace ad hoc forces with coherent structure. This elegance was not cosmetic. It was functional. It allowed gravity to be expressed in a way that respected relativistic principles and preserved consistency across scales.

The danger did not lie in the elegance itself. It lay in what the elegance encouraged people to believe.

When a mathematical structure works so well, it begins to feel complete. Each successful prediction reinforces the sense that nothing essential is missing. The theory does not merely describe outcomes; it appears to *explain* them. In such an environment, questioning foundations feels unnecessary, even disruptive. Why look beneath what already fits so perfectly?

This is how incompleteness hides behind success.

A mathematically elegant theory can be incomplete without showing any sign of failure. It can predict accurately while remaining silent about underlying mechanisms. It can organize behavior flawlessly while leaving causes untouched. Elegance masks this silence by making the description feel sufficient.

Einstein was aware of this risk. He valued simplicity and beauty, but he did not confuse them with finality. He admired elegant formulations because they reduced arbitrariness, not because they revealed ultimate reality. His discomfort with the idea that General Relativity was "finished" stemmed from this awareness. The equations were too successful to be wrong—and too silent to be complete.

The scientific community, however, often experiences elegance differently. In practice, elegance becomes a heuristic for truth. A theory that is mathematically refined is assumed to be closer to reality than one that is messy or fragmented. This assumption is not irrational; it is born of experience. Many of the most powerful theories in physics are also the most elegant.

But heuristics become hazards when they are mistaken for guarantees.

Elegance tells us that a theory is well-constructed. It does not tell us that the theory has exhausted its subject. A map can be beautifully drawn and still omit entire regions. A grammar can be flawless and still fail to name the speaker.

In the case of relativity, elegance encouraged the belief that geometry itself was fundamental. The equations related energy to curvature with such symmetry that it felt natural to treat both sides as equally real. Geometry acquired ontological weight not because it was demonstrated to be substance, but because it appeared as an indispensable partner in an elegant relationship.

This belief shaped how later problems were approached. When cosmological observations introduced tension, the response was not to ask whether the language itself had reached its limit, but to extend the language further. New terms were added. New parameters were introduced. The elegant structure was preserved by increasing its complexity.

This strategy worked mathematically. It preserved predictive success. But it also deepened dependence on a framework whose incompleteness remained unexamined. Each successful adjustment reinforced the sense that the theory was fundamentally correct, even if increasingly elaborate.

The danger of elegance lies precisely here: it discourages the question of *what is missing* by overwhelming us with how much is already explained.

Einstein avoided this trap by refusing to equate elegance with closure. His later pursuit of unification was not an attempt to replace a failed theory, but to understand why a successful one worked. He sensed that elegance at the level of consequences did not guarantee completeness at the level of reality.

Modern physics often inherits the elegance without inheriting the restraint.

There is a subtle psychological shift that occurs when elegance dominates. Instead of asking whether the theory describes reality fully, we ask how reality must behave to fit the theory. Observations are interpreted through the lens of the existing structure. Anomalies are treated as puzzles to be solved within the framework, not as invitations to rethink its scope.

This is not dogmatism. It is efficiency. It allows science to move forward without restarting from first principles at every discrepancy. But efficiency becomes rigidity when the possibility of incompleteness is forgotten.

Elegance, in this sense, acts like a seal. It closes inquiry gently, without declaring closure. It creates a sense that deeper questions are unnecessary or premature. The theory does not fail loudly. It succeeds quietly, and in doing so, it postpones reflection.

Understanding this dynamic is essential for the chapters ahead. The goal is not to undermine elegant mathematics or to devalue its role. It is to restore proportion. Elegance is a sign of internal coherence, not of ontological completeness. Precision is evidence of correct relationships, not of identified causes.

Einstein's work exemplifies this distinction. His equations are among the most elegant ever written. Their beauty reflects the coherence of the relationships they encode. But that beauty was never meant to stand in for explanation. It was meant to support it.

The power of mathematical elegance gave physics a century of progress. The danger is allowing that power to disguise the fact that something essential may still be unnamed. Recognizing this does not weaken the theory. It prepares us to extend it without illusion.

The next section restores the final piece of balance by returning to what geometry never claimed to be—and why remembering that claim matters now.

3.5 — What Geometry Never Claimed

Geometry never claimed to be reality. That claim was made later, quietly, by interpretation rather than by theory.

In Einstein's original formulation, geometry was introduced with remarkable modesty. It did not arrive announcing itself as the substance of the universe. It arrived as a disciplined way to organize relationships that had become too complex for older language. Geometry was not presented as an answer to what exists, but as a framework for how measurements relate when energy and motion are involved.

This modesty is easy to miss because it is woven into the structure of the theory rather than stated explicitly. Einstein did not write manifestos about ontology. He wrote equations that connected measurable quantities. The restraint is visible not in what he asserted, but in what he carefully avoided asserting.

At no point did Einstein claim that spacetime is a physical object. He did not say it has material properties, internal stresses, or independent agency. He did not suggest that geometry acts, pushes, stretches, or evolves by itself. These verbs entered the discussion later, largely through metaphor and convenience. Einstein himself treated geometry as descriptive structure, not as substance.

This distinction matters because geometry, by its nature, is neutral. It describes how points, paths, and intervals relate under a given set of rules. It does not specify what enforces those rules. A geometric description can apply equally to many underlying realities. That flexibility was not a weakness—it was a deliberate strength.

Einstein chose geometry precisely because it allowed him to remain silent about what lay beneath. He needed a language that could express response without committing to mechanism. Geometry offered that language. It could encode how clocks slow, how paths curve, how measurements differ, without requiring an answer to what responds or why.

This is the point where later interpretations often overreach.

When we say that mass curves spacetime, the sentence sounds like a causal explanation. But in Einstein's framework, it is shorthand for a correlation: the presence of energy is associated with a change in relational structure. The sentence does not tell us what space is, what curvature consists of, or how energy produces that change physically. It describes a consistent relationship, nothing more.

Einstein never denied that a deeper explanation might exist. He simply refused to supply one without physical grounding. His refusal was not philosophical skepticism; it was scientific discipline. To name a substance without evidence would have been to repeat the mistake of the ether—a mistake he had already witnessed and learned from.

The modesty of geometry is also visible in how little it demands. Geometry does not ask us to believe in invisible agents. It does not introduce new entities beyond what is required to describe measurement. It does not compete with matter or energy as a constituent of the universe. It sits quietly as a framework, content to do its job without insisting on interpretation.

This is why Einstein's equations can coexist with multiple worldviews. One can interpret geometry as emergent, fundamental, or effective without changing the equations themselves. The equations do not care. They operate at the level of consequence. That indifference is a sign that geometry was never meant to be ontologically loaded.

Einstein himself resisted attempts to turn his work into metaphysics. He spoke cautiously about space and time, emphasizing their relational meaning rather than their independent existence. When pressed on what spacetime "really is," he often deflected, not out of evasiveness, but out of honesty. He understood that the theory did not justify such claims.

The loss of this modesty happened gradually. As relativity became foundational, its language became familiar. Familiar language feels concrete. Metaphors hardened into intuition. Geometry, once a grammar, began to sound like a material stage. Spacetime, once a bookkeeping framework, began to resemble an actor in its own right.

This transformation was not malicious. It was the natural outcome of success combined with human cognitive habits. When a description works universally, we begin to trust it as reality. When it is elegant, we admire it. When it is indispensable, we treat it as fundamental. In this way, modest tools quietly become ontological claims without ever being examined.

Restoring the original modesty of geometry does not require rejecting any of Einstein's results. It requires remembering the limits he respected. Geometry was chosen to avoid speculation, not to replace it with a new substance. It was meant to express how systems behave, not to declare what the universe consists of.

This restoration has practical consequences. When geometry is treated as descriptive rather than substantive, questions that once seemed closed reopen naturally. We can ask what underlies relational change without contradicting relativity. We can explore the physical nature of the vacuum without dismissing geometric precision. We can seek mechanisms beneath response without abandoning the framework that describes that response so well.

Einstein left that space open intentionally. He gave physics a language powerful enough to stand on its own, but modest enough not to foreclose deeper inquiry. The later elevation of geometry to substance was not his conclusion; it was an interpretation layered on top of his work.

Understanding what geometry never claimed restores balance. It allows us to respect Einstein's achievement without inheriting assumptions he did not make. It clarifies that relativity solved the problem of describing gravitational behavior, not the problem of identifying the universe's underlying structure.

This clarity is essential for what follows. Once geometry is returned to its proper role—as grammar rather than matter—the next question emerges without confusion or confrontation. If geometry describes response, then something must be responding. Naming that responder is not a rejection of Einstein's work. It is the continuation of the honesty that defined it.

Chapter 4 — Einstein's Intellectual Restraint

"As far as the laws of mathematics refer to reality, they are not certain; and as far as they are certain, they do not refer to reality." —
Albert Einstein

The Strength to Stop Where Others Would Conclude

History often remembers bold claims more easily than deliberate restraint. It celebrates those who declare final answers and overlooks those who stop at the edge of certainty. Einstein belonged firmly to the second group. His greatest intellectual strength was not how far he pushed his ideas, but where he refused to go.

Over time, a version of Einstein emerged that he would not have recognized. In this version, he solved the universe, defined reality, and left little for others to question. This image is convenient. It turns a demanding body of work into a settled foundation. But it is also deeply inaccurate.

Einstein did not behave like a man convinced of finality. He behaved like a thinker acutely aware of limits.

He knew exactly what his equations achieved—and exactly what they did not. He was precise about predictions and cautious about interpretation. Where others saw closure, he saw an unfinished sentence. Where the community felt satisfaction, he felt unease. This unease was not doubt in his results, but doubt in their completeness.

This chapter returns to Einstein's own posture.

Not the public icon, not the simplified classroom figure, and not the metaphysical authority he was later made into. It recovers a thinker who practiced restraint as a form of rigor, who avoided naming what could not yet be grounded physically, and who resisted the temptation to let success masquerade as explanation.

Understanding this restraint matters because many of the assumptions now attributed to Einstein were never his. They emerged after him, built on his success but detached from his discipline. Recovering his stance does not weaken modern physics. It clarifies it.

Before we ask what Einstein missed, or how his work might be extended, we must first understand how carefully he chose where to stop—and why that choice was one of his most important contributions.

4.1 — Einstein's Discomfort with Finality

Einstein was never comfortable with the idea that his work represented an end point. This discomfort was not subtle, nor was it occasional. It ran through his career like a steady undercurrent, visible in his letters, his later research choices, and his repeated refusal to celebrate closure when others were eager to do so. To understand Einstein accurately, one must take this unease seriously—not as a personality trait, but as an intellectual position.

From the moment General Relativity was confirmed experimentally, the scientific community began to treat it as a completed triumph. The bending of light, the advance of Mercury's perihelion, and later confirmations all reinforced the sense that gravity had finally been "explained." Einstein, however, did not share this satisfaction in the way others expected him to. He accepted the success of the predictions, but he did not accept that the underlying problem had been settled.

This difference in attitude is crucial. Einstein did not doubt his equations. He doubted the idea that equations alone could exhaust reality.

For Einstein, a theory was complete only when it left no conceptual residue—no unanswered "why" hidden behind a working "how." General Relativity, despite its elegance, left such a residue. It described gravitational behavior flawlessly, yet it did not explain why mass–energy should alter relational structure in the first place. The equations worked, but the mechanism remained unnamed. Einstein was acutely aware of this gap.

This awareness shaped the second half of his life.

While much of the physics community moved on—toward quantum mechanics, applications, and extensions—Einstein remained fixated on unification. His pursuit of a unified field theory is often portrayed as a stubborn refusal to accept quantum mechanics or as a misguided obsession with elegance. Both interpretations miss the point. Einstein was not rejecting new physics; he was refusing to accept fragmentation as final.

To him, a universe governed by separate, irreducible principles was an admission of incompleteness. Gravity described by geometry and electromagnetism described by fields felt provisional, not fundamental. He believed that their separation indicated an unfinished understanding of nature's structure.

What is often misunderstood is that Einstein did not believe unification would overthrow relativity. He believed it would *explain* it. He wanted to know why geometry had proven to be the correct language in the first place. Why should nature respond relationally rather than mechanically? Why should energy alter the structure of motion at all? These were not questions his equations answered—and he knew it.

This is why Einstein consistently resisted the language of finality. He did not claim to have discovered the nature of space or time. He did not declare spacetime to be the ultimate substance of the universe. He spoke instead of descriptions, relations, and structures—always provisional, always open to deeper grounding.

Even his famous remarks about God and the universe reflect this stance. When Einstein spoke of order, harmony, or intelligibility, he was not asserting metaphysical doctrine. He was expressing confidence that nature is coherent beneath its appearances. That confidence did not imply closure; it implied that further understanding was possible and necessary.

The contrast with how Einstein is often portrayed could not be sharper. In popular and even academic narratives, he appears as a figure who completed a chapter of physics so decisively that all subsequent work is mere refinement. In reality, he behaved like someone who had solved a critical layer while fully expecting another to exist beneath it.

This expectation explains his resistance to probabilistic finality in quantum mechanics as well. His objections were not rooted in nostalgia for determinism, as is often claimed. They were rooted in the same discomfort with incompleteness. A theory that predicted outcomes statistically without explaining the underlying structure felt provisional to him, not wrong, but unfinished.

Einstein's discomfort with finality was therefore consistent across domains. Whether dealing with gravity, electromagnetism, or quantum phenomena, he resisted explanations that worked operationally but left structure unnamed. He accepted predictive success, but he did not confuse it with understanding.

This restraint required intellectual courage. It is far easier to accept applause than to question one's own triumph. It is far easier to let a community declare closure than to insist, quietly, that something essential is still missing. Einstein did the latter repeatedly, even when it isolated him from prevailing trends.

Einstein understood that science advances in layers. One layer can be solved completely while another remains untouched. Declaring finality at the first layer risks freezing inquiry at precisely the moment it should deepen.

This section matters because it reframes Einstein's legacy. He was not a man who believed he had explained the universe. He was a man who believed he had explained something important—and who refused to pretend it was everything. His discomfort with finality was not a flaw. It was a safeguard.

Recovering this aspect of Einstein allows us to approach his work honestly. It frees us from treating his theories as sacred endpoints and invites us to see them as disciplined milestones. It also removes a burden from modern physics: the burden of defending interpretations Einstein himself never claimed.

Einstein stopped where certainty ended. He did not close the door behind him. He left it deliberately ajar, aware that future understanding would require both the precision he achieved and the restraint he practiced.

4.2 — The Unified Field as a Clue

Einstein's later work is often described as a retreat—an old genius clinging to a lost ideal while the world moved on. This narrative is convenient, but it is deeply misleading. His pursuit of a unified field theory was not an obsession born of stubbornness. It was a clue—an intuitive response to something he knew, long before he could articulate it formally.

To Einstein, the separation of forces was never a sign of fundamental structure. It was a sign of incomplete description.

General Relativity had shown him something profound: gravity did not need to be treated as an interaction between objects. It could be understood as a relational phenomenon emerging from structure. This realization did not satisfy him; it unsettled him. If gravity could be absorbed into geometry so successfully, what did that say about the other forces? Why should nature require fundamentally different explanatory languages at its deepest level?

Einstein's intuition was simple and radical at the same time: *nature does not operate with multiple unrelated rulebooks.* Fragmentation, to him, was not richness—it was a temporary scaffolding around a deeper unity yet to be seen.

This intuition did not arise from aesthetics alone, though Einstein valued elegance. It arose from pattern recognition. He had witnessed how a conceptual shift—replacing force with response—had dissolved centuries of confusion about gravity. He suspected that similar shifts might dissolve other apparent divisions if the correct underlying structure were identified.

The unified field was not an attempt to dominate physics with a single equation. It was an attempt to identify what all interactions had in common before they were labeled as different. Einstein was not asking how forces combine. He was asking why they appear separate at all.

This distinction matters.

An obsession clings to outcomes. An intuition seeks structure.

Einstein's later work lacked experimental confirmation not because it was misguided, but because the tools required to test it did not yet exist. He was attempting to move from consequence to cause, from description to mechanism. That transition is always slower and riskier than refining predictions. The community, understandably, chose the safer path.

But Einstein's refusal to abandon unification was not a rejection of progress elsewhere. It was a recognition that progress had reached a plateau of description. The equations worked. The predictions held. The question now was not "does it calculate?" but "what makes this calculational language so universally effective?"

This is where modern portrayals go wrong. They assume Einstein was trying to replace successful theories. In reality, he was trying to explain them.

He wanted to know why geometry worked so well for gravity. Why relational descriptions appeared repeatedly across domains. Why fields, symmetries, and conservation laws kept pointing toward deeper coherence. These were not metaphysical fantasies; they were signals.

Einstein understood something that later generations often forget: *a theory that works everywhere is asking to be explained, not abandoned.*

His pursuit of unification was therefore not nostalgia for classical determinism, nor hostility toward quantum mechanics as such. His discomfort with quantum theory stemmed from the same place as his discomfort with declaring relativity complete. Both worked operationally. Both left structure unnamed.

To Einstein, probabilistic description without underlying mechanism felt provisional. Not wrong—unfinished. He was willing to accept uncertainty in outcomes, but not the idea that uncertainty itself was fundamental. That stance is frequently caricatured as philosophical stubbornness. It is better understood as methodological consistency.

The unified field, then, was Einstein's way of asking a question he had learned to respect: *what is doing the responding?*

If gravity is response to structure, and electromagnetism is response to structure, then structure itself must have a deeper physical basis. Geometry could describe it. Fields could approximate it. But neither, in his view, exhausted it.

Seen this way, Einstein's later work reads not as a failure, but as a signpost. He was pointing toward a layer of understanding that physics had not yet developed the language to articulate. He sensed that consequences were being described with extraordinary accuracy while causes remained implicit.

That sense aligns directly with the themes of this book.

Einstein did not reach the unified field, but he reached its boundary. He identified the need for it by refusing to accept that nature's coherence ends where equations succeed. His intuition was that the universe behaves as a unified whole long before it is described as one.

The tragedy of the historical narrative is that this intuition was dismissed rather than examined. The community moved on to refine consequence-models while leaving the question of unity dormant. Einstein was left appearing isolated, when in fact he was simply early.

Understanding the unified field as a clue rather than an obsession restores Einstein's intellectual continuity. His later work was not a departure from his earlier success. It was its natural extension. He had learned, through experience, that elegant description is a sign of deeper structure, not a substitute for it.

This section matters because it reframes Einstein's final decades not as a detour, but as a quiet insistence that physics had reached the edge of what consequence alone could deliver. He was not lost. He was waiting—for a language that could name what geometry had so carefully described without claiming.

That language did not emerge in his lifetime. But the clue remains.

4.3 — What Einstein Refused to Say

Einstein's intellectual discipline is revealed as much by what he avoided saying as by what he asserted. His silences were not accidental gaps; they were carefully maintained boundaries. In an era eager for grand declarations, Einstein consistently refused to cross certain lines, even when crossing them would have brought him immediate philosophical authority.

He never said that spacetime is a substance.

This is perhaps the most important refusal. Despite the widespread modern habit of speaking about spacetime as if it were a physical fabric, Einstein avoided such language. He did not describe space as something that stretches, bends, or acts in its own right. When pressed on the nature of spacetime, he redirected attention to measurement and relation. Space and time, for him, were not things but ways of ordering experience.

This refusal mattered because naming spacetime as substance would have frozen interpretation prematurely. Once something is declared to exist as a physical entity, every anomaly must be explained in its terms. Einstein did not want his theory to become a new ether—an invisible something invoked to explain everything while being observed nowhere. He had already seen how damaging that move could be.

He also refused to say that geometry causes motion.

Although later interpretations often speak as if curvature tells matter how to move, Einstein's own formulations were more cautious. Geometry, in his work, described the conditions under which motion occurs. It did not act as an agent. He avoided causal language when discussing geometric structure, even though such language would have made explanations more intuitive. He understood that intuition built on agency could easily mislead.

Einstein also refused to say that General Relativity explains gravity in an ultimate sense.

He accepted that it explains gravitational behavior. He never claimed that it explains what gravity *is*. This distinction is subtle but decisive. Explaining behavior is operational. Explaining essence is ontological. Einstein stayed firmly on the first side. When others celebrated relativity as the final word on gravity, he remained unconvinced—not because the theory failed, but because it succeeded without answering deeper questions.

Another notable refusal concerns the nature of the vacuum.

Einstein did not describe the vacuum as nothing, nor did he describe it as something with intrinsic substance. He kept the concept deliberately ambiguous. The vacuum, in his formulations, was the absence of matter and energy—not a defined physical medium with properties of its own. He allowed the equations to operate without committing to what the vacuum physically is. This restraint left space for future discovery.

He also refused to say that probability is fundamental.

Although this refusal is often discussed in the context of quantum mechanics, it reflects the same intellectual posture. Einstein did not deny the empirical success of probabilistic descriptions. He denied only the claim that probability itself represents ultimate reality. He believed that statistical behavior pointed to underlying structure not yet understood. Again, his silence was not denial—it was postponement.

Most importantly, Einstein refused to say that physics had reached completion.

Even after his greatest successes, he avoided language of finality. He did not speak of ultimate theories or finished frameworks. He did not position himself as having closed the book on fundamental understanding. This refusal was not modesty for its own sake. It was recognition that every successful description carries the risk of being mistaken for explanation.

These refusals are often overlooked because they do not fit the heroic narrative. It is easier to celebrate bold claims than disciplined restraint. Yet Einstein's authority came not from how much he claimed, but from how carefully he limited those claims.

Einstein understood that words matter. To say something in science is to constrain future thought. To refuse to say something is to preserve openness. His silences were acts of responsibility.

The scientific community, over time, filled in many of these silences on his behalf. Geometry was treated as substance. Spacetime was granted agency. Relativity was spoken of as complete. These moves were not made by Einstein, but in his name. Recovering what he refused to say helps disentangle his work from later interpretations layered onto it.

This recovery is not an exercise in historical correction alone. It has practical consequences. When we remember that Einstein never made certain claims, we free ourselves from defending them. We can revisit questions about space, time, vacuum, and causation without feeling that we are contradicting Einstein. We are, in fact, aligning with his discipline.

Einstein's refusals were not gaps in understanding. They were markers of honesty. He drew a line between what could be described and what could not yet be named. That line has since been crossed by interpretation. This section restores it—not to retreat from inquiry, but to continue it with the same restraint that defined Einstein's greatest work.

4.4 — Humility as Method

Einstein's restraint is often misread as hesitation, uncertainty, or even reluctance to commit. In reality, it was something far more rigorous. It was humility practiced as method. Not personal modesty, not philosophical doubt, but a disciplined refusal to claim more than the evidence could carry.

This distinction matters because hesitation delays action, while methodological humility enables progress without distortion. Einstein did not slow physics down by refusing to overstate his conclusions. He protected it from premature closure.

In science, there is constant pressure to explain. A theory that predicts but does not interpret fully feels incomplete. A framework that works without naming causes feels unsatisfying. The temptation is always to fill the gap—to provide a story that makes the mathematics feel real. Einstein resisted that temptation consistently, even when it would have elevated his authority beyond challenge.

Humility, in his case, was not passive. It was active boundary-setting.

He understood that every additional claim narrows future possibility. To name an entity is to commit to its properties. To declare a substance is to constrain interpretation. To assert finality is to freeze inquiry. Einstein treated these moves with suspicion, not because he feared being wrong, but because he understood the long-term cost of being prematurely right.

This is why his language remained carefully operational. He spoke in terms of measurement, relation, and response. He avoided attributing agency where none could be observed. He allowed equations to speak where words would have misled. His humility was not silence—it was precision.

Einstein practiced a form of scientific ethics that is rarely articulated but deeply influential. He separated what could be calculated from what could be claimed. He accepted success without converting it into metaphysics. He allowed his theories to stand as powerful descriptions without demanding that they be treated as ultimate truths.

This posture required confidence of a particular kind. Only someone secure in the validity of their work can afford not to overinterpret it. Einstein did not need to declare spacetime a substance to defend relativity. The predictions defended it. He did not need to insist that geometry is fundamental. The consistency of the framework spoke for itself.

Humility as method also shaped how Einstein related to uncertainty. He did not view unanswered questions as failures. He viewed them as indicators of where inquiry should slow down rather than rush forward. In an intellectual culture that often equates progress with accumulation, this stance can appear conservative. In reality, it is what allows deep progress to occur without collapse.

The contrast with later developments is instructive. As physics advanced, restraint gradually gave way to assertiveness. Interpretations hardened. Provisional language became definitive. Concepts introduced as descriptive tools were treated as elements of reality. None of this happened because evidence demanded it. It happened because success made restraint feel unnecessary.

Einstein would not have agreed with that shift. His humility was not situational; it was structural. It was built into how he thought science should operate. A theory should do its job and stop. Anything beyond that should wait until it can be grounded physically.

This is why his later dissatisfaction is so often misunderstood. It is framed as personal stubbornness or philosophical bias. In fact, it was consistency. He applied the same standard to all theories, including his own. If something worked but left structure unnamed, he treated it as incomplete—not wrong, but unfinished.

Humility as method also explains why Einstein remained open to revision without being fickle. He was not attached to specific interpretations. He was attached to coherence. If a deeper framework could explain why relativity worked, he would have welcomed it. His loyalty was not to geometry, but to understanding.

This method stands in contrast to how scientific authority is often exercised. Authority tends to close discussion. Einstein's authority, paradoxically, kept questions open. He did not use his success to silence alternatives. He used it to insist on higher standards of explanation.

Recovering this aspect of Einstein matters because it restores a lost virtue in foundational science. Humility is often treated as a personal trait—admirable, but optional. Einstein practiced it as a methodological requirement. Without it, success becomes misleading. With it, success becomes a stepping stone rather than a stopping point.

This section reframes restraint not as a lack of ambition, but as disciplined ambition. Einstein did not settle for description masquerading as explanation. He accepted description where explanation was not yet possible and refused to pretend otherwise.

That refusal was not weakness. It was the strength that kept his work from collapsing under the weight of its own success.

4.5 — How Einstein Was Later Overinterpreted

The overinterpretation of Einstein did not happen through distortion or betrayal. It happened through reverence. When a theory works as profoundly as General Relativity did, it invites completion—even when its author refused to provide one. The scientific community did not set out to misrepresent Einstein. It simply crossed lines he had deliberately drawn, often without realizing that they were lines at all.

Einstein left many doors open. Others closed them in his name.

The first crossing occurred when descriptive language was quietly upgraded to ontological claim. Geometry, which Einstein treated as a framework for expressing relationships, began to be spoken of as the substance of reality itself. Spacetime was no longer a way of organizing measurements; it became an actor. Curvature was no longer an expression of relational change; it became a causal agent. These shifts were subtle. They emerged in explanation, teaching, and metaphor long before they appeared in formal claims.

Once the language changed, intuition followed.

Students learned that mass *curves spacetime* and spacetime *tells matter how to move*. These phrases, though pedagogically useful, carried implications Einstein avoided. They suggested agency where Einstein described correlation. They implied substance where Einstein maintained neutrality. Over time, the metaphor became indistinguishable from belief.

A second crossing occurred when success was mistaken for finality.

General Relativity's predictive power created an understandable confidence. It explained phenomena that had resisted classical mechanics and did so with extraordinary precision. This success fostered a sense that gravity had been fundamentally explained, not merely described. Einstein himself never endorsed that conclusion. He celebrated the theory's correctness, not its completeness.

Yet the community increasingly spoke as if the problem of gravity was settled in principle, even if details remained. Alternative foundational questions were treated as unnecessary or misguided. Inquiry narrowed not because evidence demanded it, but because the existing framework seemed sufficient.

This shift changed how anomalies were interpreted.

When observations did not fit cleanly—whether in cosmology, galactic dynamics, or early-universe measurements—the response was not to question the scope of the framework, but to extend it. New entities were introduced. Parameters were added. Geometry was asked to do more work. None of this violated Einstein's equations, but it exceeded Einstein's restraint.

A third crossing occurred in how Einstein's philosophical posture was retroactively assigned.

Einstein's caution was often reframed as belief. His refusal to define spacetime was reinterpreted as an implicit definition. His silence about ontology was treated as endorsement of later ontological claims. In effect, Einstein was made to say things he had carefully avoided saying.

This process was reinforced by authority. Citing Einstein became a way to legitimize interpretations he never articulated. The distinction between "Einstein's equations imply" and "Einstein believed" blurred. Over time, the equations and the interpretations fused into a single inherited worldview.

Einstein himself did not participate in this fusion. He remained cautious to the end. But the momentum of scientific culture does not pause for personal restraint. As generations passed, Einstein the thinker was gradually replaced by Einstein the foundation.

This replacement had consequences.

Once Einstein was positioned as having defined reality, questioning those definitions began to feel like questioning science itself. Ontological commitments layered onto relativity acquired an air of inevitability. To ask what spacetime *is* beyond geometry felt unnecessary at best and regressive at worst. The boundary Einstein maintained was not merely crossed; it was forgotten.

The irony is that this overinterpretation did not strengthen Einstein's legacy—it narrowed it. His work became something to defend rather than something to build upon. His discipline of restraint was replaced by confidence in interpretation. The very humility that made his theories robust was omitted from their inheritance.

This section closes a necessary correction.

Einstein did not tell us what the universe is. He showed us how certain aspects of it behave with remarkable consistency. He did not define spacetime as substance. He used geometry to avoid premature substance. He did not claim finality. He warned, implicitly and explicitly, against it.

The community's overinterpretation was not malicious, but it was consequential. It shifted the center of gravity from disciplined description to assumed ontology. It made certain questions feel closed when they were never answered.

Recognizing this does not diminish Einstein. It restores him. It allows us to separate what he achieved from what was later attributed to him. It frees us to revisit foundational questions without feeling that we are undoing his work. On the contrary, it aligns us with his method.

Einstein's greatest gift to physics was not a finished picture of reality. It was a way of working that balanced precision with restraint. The overinterpretation came later, when restraint was mistaken for omission and description for declaration.

Recovering this distinction is not a critique of science. It is an act of fidelity—to the discipline Einstein practiced, not the mythology built around him.

Interlude II — When Language Becomes Reality

At some point in every mature field of knowledge, language stops describing reality and begins to *define* it. This transition is subtle. It does not arrive as an announcement or a rupture. It happens quietly, almost imperceptibly, when repeated descriptions gain the weight of certainty and metaphors harden into facts.

In science, this shift is particularly seductive. Language is not just a means of communication; it is a tool of thought. The terms we choose do more than label phenomena—they shape how we conceptualize them. Over time, descriptions that were once provisional become treated as ontological truths. What began as a way to speak about observations becomes the way reality itself is imagined.

This transformation is rarely deliberate. It arises from success.

When a model consistently predicts outcomes, its language gains authority. The metaphors embedded within it—fields, curvature, expansion, particles—begin to feel literal rather than representational. The framework stops being a map and starts being mistaken for the territory. At that point, description quietly becomes doctrine.

This is not an error of intention but a consequence of effectiveness. When a framework works exceptionally well, it resists questioning. Its metaphors solidify into assumed truths. The distinction between *how we describe* and *what is* begins to blur.

Yet language, by its nature, simplifies. It selects. It frames. It cannot capture totality; it can only gesture toward it. Every scientific term is an approximation—a handle for grasping complexity, not the thing itself. When these handles are mistaken for the reality they reference, understanding subtly narrows.

This is particularly evident in discussions of space, time, and energy. Words like "curvature," "fabric," and "expansion" are metaphors that help us visualize mathematical relationships. Over time, however, these metaphors begin to feel literal. Space bends. Time flows. The universe expands. The imagery becomes so familiar that its metaphorical origins fade from view.

At that point, language stops serving inquiry and begins to constrain it.

The danger is not in using metaphors, but in forgetting that they are metaphors. When descriptive tools are mistaken for ontological claims, questioning them feels like questioning reality itself. Alternatives are dismissed not because they are wrong, but because they sound unintuitive within the established vocabulary.

This is how conceptual rigidity forms—not through dogma, but through comfort.

The history of science shows this pattern repeatedly. Terms created to explain phenomena gradually acquire the authority of facts. Once that happens, it becomes difficult to distinguish between what is observed and how it is described. The map becomes the territory.

Recognizing this process is not an act of rejection but of refinement. It allows us to separate the usefulness of a model from its literal interpretation. It restores flexibility to thought without undermining empirical success.

In the context of this work, this distinction is crucial. The goal is not to discard established language, but to see where it has quietly overstepped its role. To ask where description has hardened into assumption, and where metaphor has been mistaken for mechanism.

Only by acknowledging this can we reopen conceptual space. Only then can new interpretations emerge without threatening the validity of what already works.

This interlude marks that pivot. It prepares the ground for moving forward—not by discarding existing frameworks, but by loosening their grip just enough to allow deeper questions to surface.

Part III — Where Description Became Assumption

Tracing the quiet shift from explanation to belief.

Chapter 5 — When Description Turned Into Belief

"Concepts that have proven useful in ordering things easily achieve such an authority over us that we forget their earthly origins and accept them as unalterable givens." — **Albert Einstein**

The Moment the Map Became the Territory

No one stood up and declared belief. No resolution was passed. No doctrine was announced.

The shift happened far more quietly.

At some point in the evolution of modern physics, descriptions that were introduced as provisional tools began to feel inevitable. Language designed to organize observation started to sound like statements about what *exists.* What began as careful phrasing gradually hardened into conviction—not through argument, but through repetition and success.

This chapter does not accuse. It traces.

Einstein's formulations were adopted because they worked. They predicted, aligned, and resolved anomalies with unmatched precision. Over time, their reliability created trust. Trust created comfort. And comfort allowed description to slip, almost unnoticed, into belief.

This was not a philosophical error. It was a human one.

When a framework explains enough, it begins to feel complete. When it feels complete, questioning it feels unnecessary. And when questioning fades, assumptions quietly take root. The language remains the same, but its role changes. What was once a way of speaking about reality becomes reality itself.

Einstein never crossed this line. Others did—not out of carelessness, but out of confidence. This chapter identifies that moment of crossing, not to assign blame, but to understand how a discipline built on skepticism can still inherit belief without realizing it.

Before we examine what followed, we must locate where the shift occurred—where description stopped being held lightly and began to be held as truth.

5.1 — From Useful Model to Assumed Reality

Interpretive drift does not announce itself as error. It presents itself as comfort.

A model is introduced to manage complexity. It organizes observations, reduces confusion, and allows prediction. At this stage, everyone involved understands—explicitly or implicitly—that the model is a tool. It stands in for something not yet fully understood. Its value lies in usefulness, not in truth claims.

The drift begins when usefulness becomes reliability.

When a model continues to work across experiments, scales, and generations of researchers, the provisional tone surrounding it softens. The language used to describe it becomes less careful. Qualifiers disappear. What was once described as "as if" slowly turns into "is." This is not because anyone decides to make that shift, but because repetition reshapes intuition.

A model that never fails begins to feel indistinguishable from reality.

This is the first stage of interpretive drift. The model stops being actively held in mind as a representation. It becomes the default mental picture. New students encounter it not as a constructed framework, but as a starting point. They are taught the conclusions before the caution. The model's origin as a response to specific problems fades into the background.

Once this happens, questioning the model no longer feels like inquiry. It feels like disruption.

The second stage follows naturally. Explanatory language begins to change. Early descriptions emphasize correlation and behavior: *this happens when that condition is present*. Later descriptions adopt causal phrasing: *this happens because that thing exists*. The model, initially descriptive, begins to carry explanatory weight it was never designed to bear.

At this point, the model has crossed an invisible threshold. It is no longer merely organizing observation; it is shaping belief.

What makes this drift particularly difficult to detect is that nothing breaks. Predictions still work. Experiments still confirm expectations. There is no empirical crisis to trigger reconsideration. The success of the model acts as a shield against scrutiny. Any unease is interpreted as philosophical discomfort rather than scientific necessity.

The third stage is institutionalization.

Textbooks simplify. Pedagogy prioritizes clarity over nuance. Metaphors are introduced to make abstract ideas intuitive. These metaphors, while helpful, carry implicit ontological suggestions. Over time, they are absorbed as literal understanding rather than illustrative aids. What began as teaching convenience becomes conceptual commitment.

Once institutionalized, the model acquires inertia. Research builds on it. Careers are shaped around it. Entire subfields depend on its assumptions. At this stage, even asking whether the model is descriptive or foundational feels misplaced. The distinction itself appears outdated.

Interpretive drift reaches completion when the model's limits are no longer visible.

At that point, anomalies are not treated as signs that the model may be incomplete. They are treated as problems that must be solved *within* the model. New parameters are introduced. Extensions are proposed. Adjustments are made. The framework stretches, but it is never questioned at the level of role.

This is not intellectual failure. It is structural momentum.

Scientific communities are not designed to constantly revisit first principles. They are designed to accumulate knowledge efficiently. Interpretive drift is the price paid for continuity. The danger arises only when continuity is mistaken for closure.

What is lost in this process is not predictive power, but conceptual humility. The original understanding—that the model was a way of speaking about reality, not a declaration of what reality is—quietly disappears. In its place emerges confidence that the language used is the language of the universe itself.

This section matters because it identifies a transition that feels natural, reasonable, and almost inevitable. No accusation is required. No error needs to be assigned. Interpretive drift is a consequence of success combined with human cognition.

Recognizing it does not weaken science. It restores awareness.

Once the distinction between model and reality is brought back into focus, questions that once seemed settled reopen without hostility. The model can continue to function exactly as before, but it no longer monopolizes interpretation. Description regains its provisional status. Belief loosens its grip.

Only from this restored posture can the next sections examine what followed—how belief shaped cosmology, how language hardened further, and why certain questions became difficult to ask without sounding unscientific.

Interpretive drift is not the moment science goes wrong. It is the moment it stops noticing how carefully it once spoke.

5.2 — Spacetime as Actor Instead of Map

The shift from spacetime as map to spacetime as actor did not require a new equation. It required only a change in how existing language was heard.

In Einstein's formulation, spacetime was a coordinate framework—a way to encode how measurements relate when energy and motion are present. It functioned as a map: precise, consistent, and indispensable. But like all maps, it was meant to represent relationships, not to perform actions. That distinction, however, proved fragile once the language surrounding it evolved.

The earliest descriptions were careful. They spoke of how matter and energy are associated with changes in relational structure. Motion followed geodesics because those paths represented the natural continuation of motion within a given configuration. There was no implication that spacetime *does* anything. It simply provides the conditions under which motion is described.

Over time, that neutrality eroded.

To make the theory intuitive, explanations began to use verbs. Spacetime bends. Spacetime tells matter how to move. Spacetime reacts. These phrases were pedagogical shortcuts, not theoretical claims. But language shapes intuition more powerfully than disclaimers. Once verbs are assigned, agency is implied. Once agency is implied, substance follows naturally in the mind.

The map had begun to move.

This shift felt harmless because it simplified explanation. Saying that spacetime acts saves time. It avoids repeated clarification about relational structure and measurement. It offers a narrative that feels complete. The problem is not that the narrative is wrong—it is that it quietly exceeds the theory's original scope.

When spacetime is treated as an actor, it becomes responsible. Phenomena are no longer described as outcomes of relational structure but as effects of spacetime's behavior. Curvature causes motion. Expansion stretches light. Geometry evolves. Each statement sounds explanatory, yet each merely restates description in active voice.

Once this active framing is normalized, deeper questions become difficult to pose. If spacetime acts, what does it consist of? If it evolves, what governs that evolution? If it responds, what is it responding with? These questions feel unnecessary only because the metaphor has already answered them implicitly.

The danger of this shift is not metaphysical speculation. It is conceptual closure.

Treating spacetime as an actor discourages inquiry into what underlies relational change. The language suggests that nothing lies beneath—that geometry itself is fundamental. This suggestion was never established by evidence. It emerged from interpretive convenience.

The transformation of spacetime from map to actor also alters how anomalies are handled. When observations challenge expectations, the actor must be adjusted. Spacetime is endowed with additional properties. New dynamics are proposed. The framework absorbs complexity rather than reflecting on its descriptive role.

This is not error. It is momentum.

Einstein never endorsed this shift. He remained cautious about attributing physical agency to spacetime. His writings emphasize relations, not actions. He understood that turning description into agent blurs the line between what is measured and what is imagined.

The community, however, found the actor-language irresistible. It offered intuitive coherence. It allowed cosmology to be narrated as a story of evolving space rather than as a record of changing relationships. It made abstract mathematics feel tangible.

But tangibility comes at a cost.

Once spacetime is treated as an actor, it competes with matter and energy for ontological status. It becomes something that exists independently rather than something that organizes measurement. This move transforms the theory from a language of consequences into a statement about substance.

The original modesty is lost.

Restoring spacetime to its role as map does not undo any prediction. It does not weaken relativity. It simply reopens a conceptual distinction that was never meant to close. Spacetime can remain the most effective language we have for describing gravitational behavior without being promoted to the thing that behaves.

This distinction matters deeply for what follows. When spacetime is an actor, explanation stops at geometry. When spacetime is a map, explanation can continue beneath it. The shift from map to actor was subtle, unintentional, and understandable—but it was a shift nonetheless.

Identifying it is not an accusation. It is an act of clarity.

5.3 — Expansion as Explanation

Expansion did not begin as an explanation. It began as a parameter.

In early cosmology, expansion was introduced to account for observed redshifts within the mathematical structure of General Relativity. It was a way of encoding how distances between reference frames change over time under certain assumptions. At this stage, expansion was not a claim about what space *is doing*. It was a way of reconciling observation with a geometric description of large-scale structure.

The interpretive shift came later.

As cosmology matured, expansion gradually moved from being a calculated outcome to being treated as a causal answer. Redshift was no longer described as *consistent with* expanding coordinates; it was described as *caused by* space itself stretching. This linguistic shift was subtle but decisive. Expansion stopped being a description of relational change and became the mechanism invoked to explain it.

Once again, nothing broke when this happened. Predictions continued to hold. Models improved. Observational precision increased. The success of expansion-based cosmology reinforced the sense that the explanation was complete. Expansion was no longer one possible interpretation among others—it became the default answer.

This default status had important consequences.

First, it inverted the direction of explanation. Instead of asking what physical processes might lead to observed redshifts and large-scale correlations, cosmology began with expansion as the given fact and worked outward. Phenomena were interpreted through it rather than examined independently. Expansion ceased to be inferred; it became assumed.

Second, expansion acquired explanatory authority beyond its original scope. It was used to account not only for redshift, but for horizon problems, structure formation, energy budgets, and temporal scaling. Each extension felt natural because the framework already enjoyed deep trust. Expansion became the unifying narrative through which disparate observations were interpreted.

Over time, this narrative hardened.

When questions arose—about discrepancies in measurements, unexpected correlations, or scale-dependent behavior—the response was rarely to question the explanatory role of expansion itself. Instead, additional components were introduced to preserve it. Dark energy, inflationary epochs, and corrective parameters emerged not as independent discoveries, but as supports required to keep expansion central.

This pattern reveals something important: expansion was no longer functioning as a description open to revision. It had become an explanatory anchor.

Once a concept reaches this status, alternatives struggle to be heard. Any competing interpretation must not only explain the data, but also justify why the default explanation should be questioned at all. The burden of proof shifts dramatically.

Expansion is treated as innocent until proven impossible, while alternatives must prove necessity.

This asymmetry is not the result of bias. It is the result of interpretive momentum.

Expansion-based explanations offered a coherent story. They allowed cosmology to be narrated as a dynamic history with a beginning, a trajectory, and an evolving structure. This narrative power made expansion more than a mathematical solution; it made it intuitive. Once intuition aligns with calculation, questioning feels unnecessary.

Yet it is precisely here that description turns into belief.

Expansion explains redshift only if one accepts that space itself is the agent of wavelength change. That acceptance is an interpretation layered onto the mathematics, not a direct consequence of it. General Relativity allows expanding solutions. It does not require expansion to be treated as the physical cause of every large-scale observation.

Einstein himself was cautious about cosmological interpretation. He explored static and dynamic models without declaring one to be ontologically privileged. His equations permitted multiple large-scale behaviors. The elevation of expansion to explanatory certainty happened after him, through accumulated consensus rather than foundational necessity.

The danger of expansion as explanation is not that it is wrong. It is that it becomes unfalsifiable in practice. When discrepancies arise, expansion can always be modified, accelerated, slowed, or supplemented. The framework absorbs challenge instead of confronting it.

This absorption gives the appearance of robustness while quietly narrowing conceptual space.

By identifying how expansion became the default answer, this section does not propose abandoning it. It proposes reclassifying it. Expansion should return to its original role: a description that fits certain observations under specific assumptions, not a universal mechanism invoked by default.

Once expansion is treated as description rather than explanation, alternative interpretations can be evaluated on equal footing. The data does not change. The posture toward it does.

This distinction is essential for understanding why certain tensions persist—not because observations are confused, but because explanation has become too comfortable to question.

5.4 — Why Questioning Slowed Down

Questioning did not slow down because curiosity disappeared. It slowed down because success made curiosity feel unnecessary.

When a scientific framework delivers consistent predictions, aligns with observation, and scales across domains, it earns trust. That trust is not blind; it is earned through repetition and verification. Over time, however, trust changes posture. What was once held cautiously becomes held confidently. What was once provisional becomes assumed. And when assumptions settle in, questioning subtly shifts from being encouraged to being inconvenient.

This is how inquiry slows—not through suppression, but through satisfaction.

In the decades following Einstein's breakthroughs, physics entered a period of extraordinary productivity. Relativity worked. Quantum mechanics worked. Technologies emerged. Predictions were confirmed. Each success reinforced the sense that the foundational structure was sound. The remaining work appeared to be refinement, not reconsideration.

Under such conditions, asking deeper questions can seem misplaced. Why revisit foundations when applications are flourishing? Why reopen conceptual debates when the machinery runs smoothly? The incentive structure of science naturally favors progress that builds forward rather than inquiry that steps sideways.

This is not intellectual laziness. It is institutional momentum.

Research funding, academic training, and publication norms reward work that operates within accepted frameworks. Young scientists are trained to solve problems, not to question the role of the tools they inherit. Conceptual restraint, which Einstein practiced instinctively, is not easily systematized. It does not produce quick results. It does not fit neatly into project proposals.

As a result, foundational questioning gradually becomes specialized, marginal, or philosophical in the pejorative sense. It is treated as commentary rather than contribution.

Success also alters the emotional relationship to theory. A framework that has survived challenge acquires a sense of inevitability. Doubting it feels not only unnecessary, but destabilizing. Questioning shifts from being an intellectual exercise to being perceived as skepticism toward science itself.

This emotional shift matters.

Once a theory is associated with the identity of a discipline, critique begins to feel personal. Alternatives are no longer evaluated neutrally; they are measured against an unspoken standard of loyalty. The question is no longer "does this interpretation explain the data?" but "why disturb what already works?"

Einstein never endorsed this posture. He understood that predictive success does not guarantee conceptual completeness. But as his work became foundational, his restraint was not transmitted alongside his equations. The community inherited the results, not the discipline that produced them.

Another factor that slowed questioning was explanatory convenience.

Frameworks like expansion-based cosmology offered narratives that felt complete. They provided beginnings, trajectories, and mechanisms that could be communicated clearly. Once such narratives are established, questioning them requires not only alternative calculations, but alternative stories. Story replacement is harder than model refinement.

As long as the narrative holds, questioning feels like overcomplication.

Moreover, the success of the framework created a feedback loop. Each anomaly was addressed by extending the existing language rather than examining its role. Adjustments preserved coherence. The framework grew more elaborate, but its position as explanatory foundation remained untouched.

This adaptive flexibility gave the appearance of openness while quietly limiting depth. Inquiry continued, but only within predefined boundaries.

The slowing of questioning was therefore not a failure of intellect, but a consequence of alignment. When theory, observation, pedagogy, and institutional practice align too well, friction disappears. And without friction, deep questioning loses urgency.

This section identifies a structural phenomenon rather than assigning blame. Science is designed to move forward efficiently, not to constantly interrogate its own foundations. Einstein was an exception because he worked at the moment when foundations were visibly unstable. Later generations inherited stability—and with it, a different set of priorities.

Recognizing why questioning slowed down is essential before attempting to revive it. The goal is not to dismantle what works, but to restore the intellectual space in which foundational questions can be asked without being treated as regressions.

Success did not end inquiry. It redirected it.

Understanding that redirection allows us to see where certain questions were postponed rather than answered—and why they now return, not as philosophical distractions, but as necessary clarifications.

5.5 — The Cost of Conflation

When language is mistaken for substance, the cost is not immediate. It accrues quietly, over time, in the form of constrained questions, layered assumptions, and explanations that feel complete while remaining structurally thin. Conflation does not break a theory. It limits what the theory is allowed to ask.

The first cost is conceptual rigidity.

Once descriptive language is treated as reality itself, the framework becomes self-sealing. Geometry is no longer a way of organizing relationships; it is what exists. Expansion is no longer a model-dependent interpretation; it is what the universe does. In such an environment, alternatives are not evaluated on explanatory merit alone—they are dismissed for violating what feels obvious.

This rigidity does not arise from dogma. It arises from familiarity. What is familiar feels real. What feels real feels unquestionable.

The second cost is explanatory inflation.

When language is promoted to substance, it must carry explanatory weight it was never designed to bear. Geometry is asked to explain causation. Expansion is asked to explain redshift, time dilation, energy balance, and cosmic history simultaneously. As the burden increases, auxiliary concepts are introduced to support the narrative.

These additions are not arbitrary. They are mathematically consistent responses to pressure. But each addition deepens reliance on the original conflation. The framework grows more complex while its foundational role remains unexamined.

The result is a system that explains everything descriptively while explaining nothing physically.

A third cost is the narrowing of falsifiability.

When description becomes belief, it becomes difficult to imagine what evidence would challenge it. Discrepancies are absorbed through modification rather than reflection. Parameters are adjusted. New epochs are proposed. The language stretches, but the assumption that the language is reality remains untouched.

This creates a paradoxical situation: a theory that is extraordinarily successful empirically but increasingly resistant to foundational challenge. Not because it is wrong, but because it has been granted explanatory privilege beyond its original scope.

The fourth cost is educational inheritance.

Students are taught conclusions without the discipline that produced them. They inherit certainty without restraint. The distinction between model and reality is rarely emphasized because it complicates instruction. Over time, generations of scientists become fluent in a language without remembering that it is a language.

This loss matters. When restraint is not taught, overinterpretation becomes invisible.

Einstein understood this risk intuitively. He treated language with caution precisely because he knew how easily it hardens into belief. His restraint protected inquiry. Its absence, inherited unknowingly, restricts it.

The final cost is philosophical without being philosophical.

Mistaking language for substance changes how reality itself is approached. Questions about what underlies behavior are replaced by confidence that behavior is all there is. Inquiry shifts from understanding to optimization. Physics becomes extraordinarily good at calculation and increasingly hesitant about meaning.

This is not a failure of science. It is the predictable outcome of success without reflection.

Conflation does not invalidate existing results. It limits their interpretive horizon. It turns provisional descriptions into assumed endpoints. It closes doors that were never meant to be closed.

This chapter ends not with correction, but with recognition.

Once the cost of conflation is seen clearly, the next step becomes possible without confrontation. Language can be returned to its proper role. Description can be respected without being worshipped. And questions that were postponed—not answered—can be asked again, not as revolutions, but as continuations.

What follows is not a rejection of what physics has built. It is an attempt to recover the clarity that once guided its greatest advances.

Chapter 6 — The Expansion Narrative and Its Limits

"A theory can be impressive and still be incomplete. Success does not certify finality." — **Albert Einstein**

When a Story Works Too Well

Cosmology did not become dominant because it was loud. It became dominant because it worked.

The expansion narrative offered something rare in science: a coherent story that connected observation, mathematics, and intuition across vast scales. Redshift found a home. Cosmic structure acquired a timeline. The universe gained a history that could be narrated from origin to present. Nothing was forced. Everything fit.

That is precisely why its limits are difficult to see.

This chapter does not dispute the achievements of modern cosmology. It does not question the legitimacy of expansion-based models or the precision of their predictions. Instead, it examines something quieter: how a successful narrative gradually became an explanatory boundary rather than a descriptive framework.

When a story works well enough, it stops feeling like a story.

Expansion began as a solution permitted by equations. It matured into a default interpretation. Eventually, it became the background assumption against which all large-scale observations are read. This shift was not an error. It was an evolution driven by success.

The purpose here is not to dismantle cosmology, but to restore proportion. To ask where the narrative explains, where it describes, and where it quietly substitutes one for the other. Only by making this distinction can the theory's strengths be preserved while its limits are acknowledged.

Before we explore what lies beyond, we must first understand how far the expansion narrative legitimately carries us—and where it begins to speak in a voice it was never meant to have.

6.1 — How Expansion Solved Many Problems

The expansion model did not rise to prominence by assumption or ideology. It earned its place by solving problems that had resisted explanation for decades. To understand why it became dominant, one must first acknowledge how much it genuinely clarified—and how effectively it organized an otherwise fragmented set of observations.

At the most basic level, expansion offered a unifying interpretation of cosmological redshift. Observations showed that light from distant galaxies was systematically shifted toward longer wavelengths. Within an expanding framework, this pattern was not merely accommodated; it became expected. Redshift could be understood as a large-scale, cumulative effect rather than a collection of local anomalies. This alone gave expansion enormous explanatory appeal.

But redshift was only the beginning.

Expansion also provided a natural context for large-scale structure. Galaxies were no longer static arrangements scattered in an eternal backdrop. They became participants in a dynamic universe with a shared history. Patterns of clustering, voids, and filaments could be discussed in terms of growth over time rather than unexplained placement. The universe acquired coherence across scale.

The model further helped resolve apparent paradoxes about uniformity. The observed isotropy of the cosmic background radiation—its remarkable sameness in all directions—found explanation in an early, dense, evolving state. Expansion allowed cosmologists to connect present uniformity to past conditions without invoking fine-tuned coincidence. The universe's large-scale smoothness could be narrated rather than merely accepted.

Expansion also brought time into cosmology in a productive way. It allowed the universe to have an age, a sequence, and a direction of development. This temporal framing transformed cosmology from descriptive astronomy into historical science. Questions could be asked not just about what exists, but about how it came to be. That shift was intellectually transformative.

Equally important was the model's mathematical compatibility. Expansion was not imposed from outside General Relativity; it emerged naturally from its equations under reasonable assumptions. This internal consistency mattered enormously. A cosmological model that arises from the same framework governing gravity and motion carried immediate legitimacy. There was no need to invent new principles at the outset.

As observations improved, the expansion framework proved flexible enough to absorb new data. It could be refined, parameterized, and tested repeatedly. Each refinement strengthened confidence. The model did not collapse under scrutiny; it sharpened.

Over time, this accumulation of success created a powerful impression: expansion was not just a useful description—it felt like the correct explanation. The model unified disparate observations into a single narrative without obvious contradiction. It made sense of scale, motion, and history in one coherent language.

This success explains why the expansion narrative came to dominate. It solved real problems. It reduced confusion. It offered continuity between theory and observation. It allowed cosmology to advance rapidly without constant foundational uncertainty.

Recognizing this matters because critique without acknowledgment is empty. The expansion model earned its authority through performance, not persuasion. Any serious examination of its limits must begin by respecting why it was trusted in the first place.

The purpose of this section is not to weaken confidence in what expansion has accomplished. It is to clarify the source of that confidence. Only then can the next question be asked honestly: whether solving many problems also allowed certain assumptions to go unquestioned.

Expansion did not dominate because alternatives were suppressed. It dominated because it worked. The challenge now is to understand whether working well is the same as explaining completely—or whether success itself can sometimes delay the next layer of inquiry.

6.2 — Redshift as Interpretation

Redshift is one of the most carefully measured and least disputed observations in modern cosmology. Light arriving from distant sources is stretched toward longer wavelengths in a systematic way.

The data is solid. The instruments are refined. The pattern is unmistakable.

What is often overlooked is that *redshift itself is not an explanation.* It is a measurement. Meaning enters only when interpretation begins.

In the early stages of cosmology, redshift was treated with appropriate caution. It was recognized as an effect requiring interpretation, not as a self-explanatory phenomenon. Multiple mechanisms were considered conceptually possible: motion, interaction, cumulative propagation effects, or large-scale structural change. Expansion emerged as the most coherent interpretation within the available theoretical framework.

Over time, that interpretation hardened.

Redshift stopped being treated as a phenomenon that *could be explained* in different ways and began to be treated as direct evidence *of* expansion itself. The distinction between observation and inference blurred. Expansion was no longer inferred from redshift; it was read directly into it.

This is where redshift began to carry more meaning than it could structurally support.

A single observable was asked to do multiple conceptual jobs at once. Redshift was used not only to indicate large-scale recession, but to justify temporal scaling, distance calibration, energy budgeting, and cosmic history. Each use was reasonable within the framework—but together they elevated redshift from indicator to cornerstone.

Once elevated, questioning its interpretation became difficult.

Redshift was no longer "consistent with expansion." It became "evidence that space itself expands." This shift matters because the latter is a stronger claim than the data alone provides. It moves from correlation to cause without an intermediate physical mechanism being identified. The observation remained the same; the narrative around it grew.

As the narrative expanded, alternative questions quietly receded. Could redshift arise from cumulative interaction with a structured medium? Could it reflect energy redistribution rather than metric stretching? Could different regimes produce similar spectral shifts through different mechanisms? These questions did not disappear because they were answered conclusively. They disappeared because redshift had already been assigned a role in a successful story.

The success of the expansion narrative amplified this effect. Because expansion explained many things at once, redshift became its most visible signature. To question the interpretation of redshift began to feel like questioning the entire cosmological framework. The weight placed on a single observable grew heavier with each success.

This is not misuse. It is overloading.

When one observation is made to support too many conceptual commitments, it becomes difficult to separate what is measured from what is assumed. Redshift itself does not tell us *why* wavelengths change. It tells us *that* they change. The explanation lives elsewhere—in theory, not in the data.

The danger of interpretive overload is not that it produces wrong answers, but that it suppresses alternative lines of inquiry. When an observation is treated as settled evidence of a specific mechanism, other mechanisms stop being explored—not because they are disproven, but because they are unnecessary within the dominant interpretation.

This is where interpretation quietly becomes belief.

The expansion framework did not invent redshift. It organized it. But organization is not identity. Redshift does not belong to expansion by necessity; it belongs to observation. The moment this distinction is forgotten, interpretation becomes indistinguishable from fact.

This section does not argue that redshift should be reinterpreted immediately or replaced wholesale. It argues that *redshift was asked to carry ontological meaning it was never measured to contain*. That burden emerged from narrative success, not from observational mandate.

Recognizing this restores proportionality. Redshift can remain one of the strongest observational pillars in cosmology without being treated as a universal explanatory key. Once its role is properly bounded, cosmology gains flexibility rather than losing coherence.

Only then can limits be examined without confrontation—and only then can alternative interpretations be evaluated on their merits rather than dismissed by association.

6.3 — Scale, Consistency, and Convenience

The expansion narrative did not fail at scale. It succeeded so well across scales that its internal tensions were easy to overlook. What eventually appeared as cracks were not dramatic contradictions, but quiet mismatches—subtle signals that consistency was being preserved through convenience rather than through deeper understanding.

At small and intermediate scales, expansion-based reasoning works smoothly. Distances can be parameterized, redshifts correlated, and motions approximated without difficulty. The framework behaves predictably where assumptions remain close to the conditions under which it was calibrated. This local consistency reinforced confidence that the same logic could be extended universally.

Trouble emerged only when scale itself became the subject.

As observations reached farther back in time and farther out in distance, the framework was asked to operate simultaneously across regimes that were never directly accessible. Early-universe conditions, late-universe measurements, and local structures were all interpreted through the same expansion lens, even though the physical environments they represent differ radically.

To maintain consistency, cosmology leaned on convenience.

Parameters were assumed to be transferable across epochs. Relationships calibrated locally were extrapolated globally. Smoothness was imposed where direct verification was impossible. These moves were not unreasonable—they were necessary to make progress. But necessity and truth are not the same thing.

The first quiet crack appeared in calibration.

Measurements that depended on local astrophysical objects began to diverge from those inferred from early-universe signals. The discrepancy was not large enough to break the model, but persistent enough to resist elimination. Rather than prompting reconsideration of interpretation, it was treated as a technical problem to be resolved within the existing framework.

This response revealed something important.

The framework had reached a point where internal consistency was prioritized over interpretive flexibility. If results disagreed, the assumption was not that the narrative might be incomplete, but that some piece of the machinery needed refinement. Convenience dictated that the structure be preserved while adjustments occurred at the edges.

A second crack appeared in the handling of uniformity.

Expansion-based cosmology relies heavily on large-scale smoothness. This assumption simplifies equations and aligns with background observations. Yet the actual universe is profoundly structured. Clusters, voids, filaments, and local anisotropies dominate most observational contexts. Bridging the gap between smooth global models and lumpy local reality required increasingly sophisticated averaging procedures.

These procedures worked mathematically. Conceptually, they introduced ambiguity.

At what scale does smoothness meaningfully apply? How does local structure influence global interpretation? When does averaging conceal rather than clarify physical behavior? These questions lingered without disrupting calculation, but they indicated that consistency was being managed rather than derived.

The third crack appeared in explanatory layering.

As new observational tensions arose, solutions tended to be additive. Additional components were introduced to preserve the expansion narrative rather than to question its explanatory role. Each addition resolved a specific issue, but the overall structure became more dependent on assumptions that could not be independently verified.

Convenience again played a role.

It was far easier to extend a familiar framework than to revisit its foundations. The mathematics supported extension. The community understood the language. The narrative remained intact. From a practical standpoint, this was efficient. From a conceptual standpoint, it postponed reflection.

What makes these cracks "quiet" is that none of them invalidate the model. The equations still work. Predictions remain accurate within uncertainty. No single observation demands collapse. But taken together, they suggest that consistency is being maintained through careful tuning rather than through a fully grounded physical picture.

This is not failure. It is a signal.

When a framework depends increasingly on convenience to preserve coherence, it indicates that description may be outrunning explanation. The model is doing more organizational work than interpretive work. It is holding data together without necessarily revealing what holds reality together.

Scale plays a crucial role here. At certain scales, the assumptions feel natural. At others, they feel imposed. The same language is stretched across regimes where its physical meaning becomes less clear, even as its mathematical form remains stable.

This section does not argue that the expansion narrative is wrong. It argues that its limits become visible only when scale, consistency, and convenience are examined together. The cracks do not appear as breaks. They appear as dependencies—on calibration, averaging, and extension.

Recognizing these dependencies does not weaken cosmology. It contextualizes it.

Once convenience is acknowledged as a practical necessity rather than an ontological guarantee, the framework can be appreciated for what it achieves without being burdened with claims it was never designed to support. The next section explores how those limits became difficult to articulate precisely because success made them easy to ignore.

6.4 — Dark Components as Adjustments

Dark Matter and Dark Energy did not enter cosmology as confessions of failure. They entered as acts of preservation.

When observations strained the expansion framework, the response was not denial. It was accommodation. Additional components were introduced to protect internal consistency while allowing the narrative to continue doing its explanatory work. In this sense, the dark components functioned as structural necessities rather than signs that something had gone wrong.

This distinction matters.

Dark Matter was introduced to reconcile gravitational behavior with visible mass. Galaxies rotated too quickly. Clusters held together too strongly. Within an expansion-based, geometry-centered framework, the simplest way to preserve consistency was to assume additional, unseen mass. The move was mathematically sound. It restored balance to equations without altering the underlying narrative.

Dark Energy followed a similar path.

As observations suggested accelerated expansion, the framework again faced a choice: question the role of expansion itself or introduce a compensating term. The latter option preserved continuity. A new component was added to account for large-scale behavior without revisiting the interpretive foundation. Acceleration could now be described without redefining what expansion meant.

In both cases, the additions worked.

They stabilized models. They aligned predictions with data. They allowed cosmology to remain internally coherent while absorbing new information. This is why dark components gained acceptance quickly. They were not arbitrary patches; they were structurally elegant solutions within the existing language.

But elegance within a language does not imply necessity in reality.

The critical point is not whether dark components exist. It is why they were required. They arose because the descriptive framework was asked to explain more than it originally promised. Geometry and expansion were elevated from organizing principles to causal explanations. Once that elevation occurred, any discrepancy had to be resolved *within* that elevated role.

Dark components became the means of preserving that role.

This framing reveals why dark matter and dark energy are often described as accounting for most of the universe's content. They do not merely supplement visible matter; they uphold the explanatory structure itself. Without them, the narrative fractures. With them, it holds.

Seen this way, dark components are not failures of observation. They are indicators of interpretive load.

They signal that the framework is doing essential organizational work while lacking a fully grounded physical account of what organizes it. Rather than prompting foundational reconsideration, the response was to add terms that allowed the description to remain intact.

This strategy is rational. It is how science often progresses. When a framework works broadly, it is extended rather than replaced. But extension has limits. Each added component increases dependence on assumptions that cannot be directly tested in isolation.

Over time, this creates a peculiar situation: a universe whose dominant constituents are defined entirely by what they do not do.

They do not emit light. They do not interact electromagnetically. They do not behave like known matter. They exist because they are required for consistency, not because they are observed directly.

This does not make them fictional. It makes them functional.

The cost of this functionality is interpretive inertia. Once dark components become embedded, questioning the framework that required them becomes harder. Any alternative interpretation must now explain not only the original observations, but also the success of the adjustments.

This section does not argue against dark matter or dark energy as possibilities. It reframes their role. They are best understood as structural responses to pressure within a successful descriptive system, not as definitive revelations about the universe's composition.

Understanding this reframing allows cosmology to retain its achievements without mistaking necessity for explanation. Dark components preserve coherence. They do not, by themselves, complete understanding.

The next section explores why acknowledging this distinction does not weaken cosmology—but instead clarifies where its narrative legitimately ends and where further inquiry must begin.

6.5 — A Model Under Strain, Not Collapse

The expansion narrative is not collapsing. It is straining—and that distinction is essential.

Collapse implies failure, contradiction, or loss of predictive power. None of these apply. Expansion-based cosmology continues to describe observations with remarkable accuracy. Its equations remain coherent. Its models remain testable. Its successes far outweigh its uncertainties. To frame the current moment as crisis would be misleading.

Strain, however, is something else.

Strain appears when a framework is asked to do more interpretive work than it was originally designed to handle. It shows up not as error, but as tension—between scales, between measurements, between descriptive convenience and physical clarity. The expansion model is under strain because it is being stretched across domains where its foundational assumptions are increasingly visible.

One source of strain lies in consistency across regimes.

Early-universe signals and late-universe measurements do not always align cleanly. Local structures complicate global averages. Calibration methods depend on layered assumptions. None of this invalidates the framework, but it requires increasingly careful management. The system holds together, but only with sustained effort.

Another source lies in interpretive accumulation.

Expansion now carries explanatory weight for redshift, cosmic history, energy distribution, and temporal scaling. Each role is defensible in isolation. Together, they place heavy demands on a single interpretive concept. The framework absorbs these demands through additional components and refinements, but the cumulative load is nontrivial.

A third source lies in conceptual clarity.

As the model becomes more elaborate, the distinction between description and explanation grows harder to maintain. Expansion describes relational change effectively. Whether it explains the physical origin of that change remains less clear. This ambiguity does not stop progress, but it marks a boundary the framework approaches repeatedly without crossing.

Importantly, none of these strains suggest abandonment.

Scientific models are not judged solely by whether they strain. They are judged by whether they break. The expansion narrative has not broken. It continues to guide research productively. But acknowledging strain allows for intellectual honesty without antagonism.

This acknowledgment also prevents defensive escalation.

When strain is mischaracterized as threat, the response is often to entrench. When it is recognized as a natural consequence of success, the response can be exploration. The expansion model can be treated as a powerful description that may one day be embedded within a deeper account, rather than as a final explanation that must defend itself indefinitely.

Einstein understood this posture instinctively. He never interpreted strain as failure. He saw it as signal. His own work strained classical mechanics before replacing it. Relativity itself strained Newtonian intuition long before it was accepted. Strain is not the end of a model's life; it is often the beginning of its next role.

Seen this way, the current state of cosmology is not precarious. It is transitional.

The expansion narrative has reached a level of success that exposes its own assumptions. That exposure is not an accusation. It is an opportunity. It invites careful distinction between what the model does exceptionally well and what it was never meant to declare.

This section closes by restoring balance. Expansion remains a valid, productive, and necessary framework. Recognizing its strain does not weaken it. It places it in context.

What follows is not rejection, but refinement of posture. The next chapter asks what becomes possible when expansion is treated once again as description rather than destiny—when the narrative is respected without being mistaken for the thing it describes.

Chapter 7 — The Hubble Tension: A Message Misread

"The important thing is not to stop questioning. Curiosity has its own reason for existing." — **Albert Einstein**

When Disagreement Is Not a Problem but a Signal

The Hubble tension is often spoken of as a problem to be solved. A discrepancy to be eliminated. An error to be corrected.

But what if it is none of these?

What if the tension is not a malfunction in measurement, but a message carried by the measurements themselves? Not a contradiction, but a signal that the story we are telling has reached the edge of its usefulness.

The disagreement between early-universe and late-universe measurements of cosmic expansion is precise, persistent, and remarkably stable. It refuses to vanish with better instruments or refined statistics. Instead, it sharpens. That alone should provoke curiosity rather than concern.

Yet much of the response has focused on reconciliation—finding adjustments that will make the numbers agree again. New parameters are introduced. Systematics are reexamined. Calibration methods are refined. Each effort is reasonable. Together, they reveal something deeper: an instinct to preserve a narrative rather than to question its premises.

This chapter proposes a different posture.

What if the Hubble tension is not a flaw in measurement, but a boundary marker? What if it is not a problem to be solved, but a signal pointing beyond the framework that generated it? In that case, the tension is not an obstacle—it is information.

Throughout scientific history, moments like this have marked transitions. Discrepancies that refused to vanish eventually led to deeper understanding, not by being forced into existing models, but by exposing the limits of those models. The Hubble tension may be such a moment.

This chapter does not argue that current cosmology is wrong. It suggests that it may be incomplete in a specific and instructive way. The disagreement between early and late measurements is not noise to be eliminated, but structure waiting to be interpreted.

If earlier chapters traced how description became assumption, this chapter examines what happens when assumptions begin to strain. The question is no longer whether expansion works—it clearly does—but whether it tells the whole story.

Here, the tension itself becomes the message.

7.1 — What the Hubble Tension Really Is

At its simplest, the Hubble tension is a disagreement about how fast the universe is expanding. Not in theory, but in measurement. Two well-tested ways of measuring the same thing—how quickly space appears to be stretching—give two different answers. Both are precise. Both are reliable. And yet, they do not agree.

This alone makes the issue important.

On one side are measurements that look far back in time, observing the early universe through relic radiation left over from its infancy. These measurements provide a value for the expansion rate based on conditions when the universe was young, dense, and uniform. On the other side are measurements made in the relatively nearby universe, using stars and galaxies whose distances can be measured more directly. These give a higher value for the same quantity.

The disagreement is small in absolute terms, but large in meaning. It is not random error. It persists across different instruments, methods, and research groups. As measurement precision improves, the gap does not close. It becomes clearer.

This is what makes the situation unusual.

If one set of measurements were flawed, the discrepancy would fade with better data. If assumptions were wrong, corrections would converge. Instead, each refinement sharpens the difference. The universe appears to be telling two consistent but incompatible stories about its rate of expansion.

This is why the issue is called a "tension." Not because the data are unreliable, but because they are too reliable to ignore.

At its core, the tension arises from comparing two different kinds of observation. One reconstructs the universe's behavior from its early imprint. The other measures its current state directly. Both approaches are valid. Both are grounded in well-tested physics. Yet they do not align.

This does not mean one is wrong and the other right. It suggests that the assumptions connecting them may be incomplete.

The common response has been to search for hidden errors, unknown biases, or missing parameters. Perhaps early-universe physics was slightly different. Perhaps late-time measurements are affected by subtle astrophysical effects. These are reasonable avenues of investigation. But they share a common assumption: that the underlying framework is sound, and only its details need adjustment.

What if that assumption itself deserves scrutiny?

The Hubble tension may not be a problem within the model but a signpost at its boundary. It may indicate that the way expansion is conceptualized—how time, distance, and energy are related across cosmic history—requires refinement.

Seen this way, the tension is not a flaw to be corrected but information to be interpreted. It signals that the universe is not behaving inconsistently, but that our description may be incomplete.

This reframing changes the tone of the discussion. Instead of asking, "Which measurement is wrong?" we can ask, "What assumption connects these measurements, and is that assumption universally valid?"

The power of the Hubble tension lies precisely in its persistence. It resists resolution because it is not a technical glitch. It is a structural hint. It tells us that something subtle is being overlooked—not something dramatic, but something foundational.

In this sense, the tension is not a problem to eliminate. It is a message waiting to be read.

7.2 — Why Data Is Not the Problem

One of the most important clarifications in the discussion around the Hubble tension is also the simplest: the data itself is not failing. The measurements that produce the discrepancy are internally consistent, repeatedly verified, and obtained through independent methods. The problem does not lie in faulty instruments or careless observation. It lies elsewhere.

This distinction matters because the instinctive response to disagreement in science is often to suspect error. When two measurements do not align, the natural assumption is that one—or both—must be wrong. But in the case of the Hubble tension, that explanation has steadily lost credibility. Each method involved has undergone years of refinement, cross-checking, and validation. The disagreement has persisted precisely because the measurements are reliable.

One side of the tension comes from observations of the early universe. These measurements rely on well-understood physical processes that occurred shortly after the universe became transparent to radiation. The data extracted from this epoch is remarkably clean, governed by linear physics and constrained by well-tested models. The internal consistency of these observations is extraordinarily high.

The other side comes from measurements of the nearby universe—direct observations of cosmic distances using multiple independent techniques. These measurements, too, have been refined repeatedly. Corrections for bias, calibration errors, and environmental effects have been applied with increasing precision. The result is not convergence, but stability around a value that remains distinct from the early-universe estimate.

If either side were poorly understood, the discrepancy might be dismissed. But both sets of measurements are internally coherent. They agree within themselves even as they disagree with each other. That is what elevates the tension from a technical nuisance to a conceptual problem.

This distinction is critical. When data sets are inconsistent internally, the solution is methodological improvement. When they are internally consistent but mutually incompatible, the issue points elsewhere—toward assumptions that bridge them.

The persistence of the Hubble tension despite improved data suggests that the disagreement is not about measurement quality but about interpretation. The same physical quantity—cosmic expansion rate—is being inferred through different pathways, and those pathways do not converge. This indicates that the underlying model linking those pathways may be incomplete.

It is tempting to resolve such tensions by refining parameters or introducing additional variables. These approaches are often productive. But they also risk obscuring the deeper question: whether the framework connecting early- and late-universe measurements is fully capturing the relevant physics.

The tension therefore acts as a diagnostic. It highlights a boundary where existing assumptions may no longer apply uniformly. Rather than indicating error, it signals sensitivity—a point where the model's internal coherence is being tested.

Importantly, this does not undermine the credibility of observational cosmology. On the contrary, it demonstrates its maturity. Only a field with exceptionally precise measurements can encounter such a problem. The data are not contradicting each other; they are speaking from different regimes within the same theoretical structure.

Recognizing this shifts the conversation. Instead of asking which measurement is wrong, the more productive question becomes: what assumption connects them, and why might that assumption fail across scales?

This perspective reframes the Hubble tension from a problem to be fixed into a clue to be interpreted. It suggests that the universe is not misbehaving; it is communicating that our unifying narrative may be incomplete.

Understanding this distinction is crucial before any deeper interpretation can begin. Without it, discussions risk circling endlessly around calibration and correction. With it, the tension becomes what it truly is: a signal pointing beyond the limits of our current descriptive framework.

7.3 — Interpretation Under Pressure

When observations refuse to align, pressure does not fall on the data—it falls on interpretation. This is the stage where a framework is tested not by what it predicts, but by how it reacts when prediction becomes strained. In the case of the Hubble tension, this pressure has revealed the boundaries of the interpretive assumptions underlying modern cosmology.

At first, the response was technical. Researchers examined calibration methods, refined distance ladders, improved instrumentation, and rechecked statistical treatments. These efforts were necessary and valuable. But as refinements accumulated without resolving the discrepancy, the nature of the problem quietly shifted. The issue was no longer precision; it was interpretation.

The core assumption under strain is simple: that the same physical description applies uniformly across cosmic history. This assumption allows early-universe measurements and late-universe observations to be compared directly, as though they were snapshots of the same system taken at different times. When this assumption holds, discrepancies signal error. When it fails, discrepancies signal structure.

What makes the current situation unusual is that both sides of the measurement divide are internally consistent. Each tells a coherent story within its own domain. The tension arises not from noise, but from the attempt to force these stories into a single explanatory narrative without adjusting the underlying assumptions that connect them.

This is where interpretation becomes critical.

If expansion is treated as a uniform, global process acting identically across all scales and epochs, then disagreement between measurements appears pathological. But if expansion is instead understood as a contextual description—one that may manifest differently depending on scale, environment, or regime—then the tension becomes intelligible rather than alarming.

The difficulty lies in recognizing when a model has shifted from explanation to assumption.

For decades, cosmology has operated under the premise that expansion is a universal descriptor. That premise worked extraordinarily well within certain domains. However, as observational reach extended both backward and outward, the assumption began to stretch. The model was being asked to account for phenomena far removed from the conditions under which its core relationships were established.

This is where interpretive pressure accumulates. Instead of re-examining the scope of the framework, refinements are layered atop it. Additional parameters, corrections, and auxiliary hypotheses are introduced to preserve continuity. Each addition is rational. Collectively, they indicate strain.

The problem is not that these additions are wrong, but that they reveal a growing distance between description and foundation. The framework continues to function, but increasingly as a system of adjustments rather than as a unified explanation.

In this sense, the Hubble tension is not an anomaly within the data. It is an anomaly in the alignment between description and reality. It marks the point where interpretation is doing more work than observation.

This does not invalidate the existing model. It highlights its limits.

All scientific frameworks reach a stage where refinement alone cannot resolve emerging discrepancies. At that point, progress requires re-examining the assumptions that were previously invisible because they worked so well.

The pressure revealed by the Hubble tension is not a threat to cosmology. It is an invitation. It suggests that the current conceptual apparatus, while extraordinarily successful, may be approaching the edge of its descriptive reach.

Recognizing this does not weaken the science. It strengthens it by opening space for deeper questions—questions that cannot be answered by adding terms or adjusting parameters, but only by reconsidering what the model is truly describing.

In this sense, the tension is not a failure of interpretation, but a signal that interpretation has reached its limit and must evolve.

7.4 — When a Constant Refuses to Be Constant

There are few things more unsettling to a scientific framework than a constant that refuses to behave like one. Constants are meant to anchor understanding. They are the quiet assurances that allow models to stand upright. When a value called "constant" begins to vary—depending on how or where it is measured—it does more than introduce uncertainty. It calls the structure of interpretation into question.

The Hubble constant occupies exactly this uneasy position.

It was named a constant not because it was proven immutable, but because it was assumed to represent a universal property of the cosmos: a single rate characterizing expansion everywhere and at all times. This assumption made sense within the early logic of cosmology. A uniform universe expanding uniformly would naturally possess a single defining rate. The elegance of this idea made it compelling.

But as measurement techniques improved, the constancy of the constant began to erode.

Different methods, applied to different epochs and scales, produced different values. The divergence was not random noise; it was systematic. The more carefully each method was refined, the more stubborn the disagreement became. What should have converged instead stabilized into a split.

This is the moment when a parameter stops behaving like a constant and starts behaving like a clue.

At first, the response was to search for error. Calibration issues, hidden biases, unaccounted variables—every reasonable possibility was explored. When these avenues failed to resolve the discrepancy, the tension was reclassified as "interesting but unresolved." The label softened the discomfort without addressing its cause.

Yet a constant that varies is not merely inconvenient. It challenges the conceptual scaffolding that required it to be constant in the first place.

If the expansion rate depends on when and how it is measured, then expansion itself may not be the uniform, global phenomenon it was assumed to be. It may instead reflect a more complex process—one whose behavior depends on context, structure, or scale. In that case, the idea of a single number governing cosmic growth becomes a simplification rather than a truth.

This does not imply that the universe is chaotic or incoherent. It suggests that the language used to describe its large-scale behavior may be too coarse. Just as temperature loses meaning at very small scales, expansion may lose its universality when examined too closely.

The discomfort arises because constants are supposed to stabilize our theories. They anchor equations and allow comparison across time and space. When a constant becomes conditional, it forces a re-evaluation of what is fundamental and what is emergent.

This is where interpretation matters.

One can choose to preserve the idea of a universal expansion rate by introducing additional entities or mechanisms that reconcile the discrepancy. Or one can treat the discrepancy as a sign that the expansion framework itself is a higher-level approximation—accurate within a range, but not fundamental.

The second option is less comfortable because it shifts the question from "what parameter is wrong?" to "what assumption is misplaced?" It requires rethinking what the constant was meant to represent in the first place.

A constant that varies is not necessarily broken. It may be telling us that what we thought was fundamental is actually emergent—arising from deeper dynamics that are not captured by the current description.

Seen in this light, the Hubble tension is not a technical nuisance. It is a philosophical signal embedded in data. It tells us that our conceptual scaffolding, not our measurements, may be due for revision.

The resistance to this idea is understandable. Constants provide stability. They anchor equations and anchor confidence. Letting go of their absoluteness feels like stepping into uncertainty.

Yet science advances precisely by recognizing when constants are not as constant as assumed.

In this sense, the tension is not an anomaly to be erased but an invitation to rethink what constancy itself means. It challenges the assumption that a single number can encapsulate the dynamic behavior of an evolving universe.

The discomfort it generates is not a sign of failure. It is a sign that understanding is pressing against its current boundaries, asking for a broader frame in which stability and change can coexist.

7.5 — Tension as Signal

Tension is often treated as a problem to be eliminated. In science, as in engineering, inconsistency is usually read as error, something to be smoothed out or corrected. But not all tension signals failure. Some forms of tension are diagnostic. They indicate not breakdown, but boundary.

The Hubble tension belongs to this second category.

It is not the kind of discrepancy that grows chaotic or unstable. It does not explode across datasets or collapse under refinement. Instead, it remains remarkably stable. Independent methods converge internally while disagreeing with one another in a structured way. This persistence is not noise—it is information.

When a system produces consistent results that disagree across frameworks, the implication is not that the data are wrong, but that the framing is incomplete.

In this sense, the tension acts like a stress test. It applies pressure to the assumptions that connect observation to interpretation. Under that pressure, some assumptions hold, while others begin to show limits. This is not failure; it is diagnostic clarity.

Historically, the most productive moments in physics have arrived not when contradictions vanished, but when they refused to disappear. The orbit of Mercury, blackbody radiation, and the photoelectric effect all appeared first as anomalies—small tensions within otherwise successful theories. Each pointed beyond the framework that generated it.

The current tension plays a similar role.

It does not demand the abandonment of cosmology. It asks for reflection on what cosmology presumes. It suggests that the relationship between observation and explanation may not be as linear as once assumed. The discrepancy invites a deeper look at how quantities are defined, what assumptions link them, and where those assumptions quietly shape interpretation.

Crucially, the tension does not undermine the validity of the data. Instead, it elevates it. Only robust data can sustain such a persistent inconsistency. If measurements were weak or unreliable, the tension would dissolve under scrutiny. That it does not is precisely what makes it informative.

Seen this way, the tension becomes a guidepost rather than an obstacle. It points to the boundary of the current framework—the edge beyond which existing concepts begin to blur. It suggests that the difficulty lies not in measurement, but in the conceptual lens through which measurements are combined.

This reframing changes the tone of the discussion. Instead of asking, "Which value is correct?" the more productive question becomes, "What assumption allows these values to coexist?" The focus shifts from reconciliation to understanding.

Such a shift requires intellectual patience. It resists the impulse to resolve tension immediately. It treats inconsistency not as a flaw but as a message. And it invites reconsideration of foundational ideas that have long gone unquestioned.

The history of science shows that its deepest advances often begin this way—not with contradictions that destroy theories, but with tensions that reveal their boundaries. The present moment fits that pattern.

The Hubble tension does not demand resolution through adjustment. It invites interpretation. It asks not for new parameters, but for a reexamination of what parameters mean.

Seen in this light, the tension is not an embarrassment to be explained away. It is a signal—quiet, persistent, and precise—that something fundamental remains to be understood.

And that is not a failure of science. It is its invitation forward.

Interlude III — When Consistency Becomes Constraint

Consistency is one of the great virtues of scientific thought. It is what allows ideas to endure, theories to cohere, and knowledge to accumulate across generations. Without consistency, science would fracture into disconnected insights, unable to build upon itself. And yet, there comes a moment—quiet, often unnoticed—when consistency begins to limit rather than liberate.

This moment does not announce itself as failure. On the contrary, it arrives wrapped in success. A theory works so well, explains so much, and predicts so reliably that its internal logic becomes unquestioned. Its coherence becomes its authority. What began as a framework becomes a boundary.

At this point, consistency transforms from a guide into a constraint.

The problem is not that the framework is wrong. It is that it has become complete in its own terms. Every new observation is interpreted within its structure. Every anomaly is adjusted, absorbed, or reinterpreted until it fits. The theory does not break—it adapts. But in doing so, it may also begin to close off alternative ways of seeing.

This is not a failure of rigor; it is a consequence of success.

A consistent framework is stable. Stability encourages trust. Trust discourages questioning. Over time, the distinction between explanation and assumption fades. The framework becomes self-validating, not because it explains everything, but because everything is explained *through* it.

At this stage, anomalies no longer threaten the structure. They are absorbed, reclassified, or deferred. The theory survives, but inquiry narrows. What once invited exploration now enforces coherence.

This is the subtle danger of conceptual maturity.

When a framework reaches this level of internal harmony, it can become difficult to ask foundational questions without appearing regressive or unproductive. The very tools that enabled progress begin to define the limits of what can be thought. Exploration gives way to refinement. Interpretation replaces curiosity.

Yet history shows that the deepest shifts in understanding occur precisely when this kind of stability is questioned—not rejected, but gently unsettled. When thinkers pause to ask not whether a theory works, but whether it still speaks to the whole of what it seeks to explain.

Such moments are rare because they require intellectual humility. They require the willingness to consider that success may conceal incompleteness, and that consistency may mask deeper tensions.

This interlude marks that moment of pause.

It does not argue that existing frameworks are wrong. It suggests they may be complete only within their own assumptions. It invites reflection on whether those assumptions still serve the questions we now face.

When consistency becomes constraint, the path forward is not destruction but reorientation. It is the recognition that explanation has limits, and that beyond those limits lies not confusion, but the possibility of a more encompassing view.

The purpose of this pause is not to undermine confidence, but to restore openness. To remind us that every great advance in understanding has come from recognizing where explanation ends and inquiry must begin again.

In that space—between what we know and what we assume—we find the conditions for genuine discovery.

Part IV — Reopening The Question Of Reality

From description toward deeper structure.

Chapter 8 — The Vacuum That Was Treated as Nothing

"What we observe is not nature itself, but nature exposed to our method of questioning." — **Werner Heisenberg**

The Absence That Was Never Empty

For most of modern physics, the vacuum has been treated as absence—an empty stage on which the universe performs. It was defined not by what it is, but by what it lacks. No matter. No radiation. No structure. A background so neutral it could be ignored.

And yet, this "nothing" has always behaved like something.

It bends light. It transmits forces. It stores energy. It responds to boundaries. It fluctuates, resists, and participates—quietly but persistently. The more carefully physicists looked, the less empty it became.

Still, the language remained unchanged.

We continued to speak of vacuum as if it were a passive backdrop, even as evidence accumulated that it plays an active role in shaping physical behavior. The contradiction was tolerated because the mathematics worked. The language, though strained, held together well enough to proceed.

This chapter begins where that habit begins to fray.

The vacuum was never nothing. It was simply undefined. Not empty, but unarticulated. Treated as absence because there was no conceptual framework for its presence. The absence was assumed, not demonstrated.

By labeling the vacuum as "nothing," physics avoided a deeper question: *What kind of thing must exist for fields, forces, and energy to propagate at all?* The question was deferred, not answered.

That deferral was practical. It allowed progress. But it also created a blind spot.

As observational precision increased and theoretical structures grew more intricate, the vacuum began to appear less like a void and more like a participant. Energy fluctuations, zero-point effects, and vacuum polarization all hinted at an underlying structure that could not be dismissed as mere emptiness.

Yet language lagged behind observation. The vacuum remained linguistically empty even as it grew physically active.

This chapter does not claim to define what the vacuum is. It does something quieter and more necessary. It asks whether calling it "nothing" has prevented us from seeing what it does.

By examining how the vacuum was framed, and how that framing shaped interpretation, we can begin to see how an unexamined assumption became foundational. Not because it was proven, but because it was convenient.

To move forward, the vacuum must be reconsidered—not as a mystical substance, nor as a metaphysical placeholder, but as a structural participant whose role has been underestimated.

This chapter begins that reconsideration.

8.1 — The Vacuum as Absence

For most of scientific history, the vacuum has been defined not by what it is, but by what it lacks. It was introduced as an absence—of matter, of substance, of influence. This definition was not born from discovery but from necessity. In order to describe motion, force, and interaction, physics required a backdrop that did nothing. The vacuum became that backdrop: an empty stage upon which physical events could unfold without interference.

This conception was not arbitrary. It emerged naturally from classical mechanics, where space functioned as a passive arena and matter as the active participant. When forces acted, they acted *within* space, not *through* it. The idea that the vacuum itself might possess properties was unnecessary and, for a long time, unimaginable. Simplicity demanded emptiness.

Even as physics evolved, this assumption persisted. The vacuum was refined, redefined, and mathematically reworked, but its conceptual status remained largely unchanged. It was treated as a baseline—a neutral reference against which motion, energy, and structure could be measured. It was not expected to contribute anything of its own.

This inherited view shaped how generations of scientists thought about the cosmos. If space was empty, then any observed effect had to originate from matter or fields placed within it. The vacuum was not a participant; it was a container. This assumption allowed theories to remain clean and calculations tractable.

Yet this simplicity came at a cost.

As physics progressed into regimes of extreme scale and energy, the notion of an inert vacuum became increasingly strained. Phenomena began to appear that could not be neatly attributed to matter alone. Fluctuations emerged where nothing was supposed to be. Forces seemed to arise without sources. The "emptiness" behaved in ways that contradicted its definition.

Rather than abandoning the concept, physics adapted around it. The vacuum was redefined incrementally—still called empty, yet endowed with properties. Zero-point energy, virtual particles, vacuum polarization: these ideas preserved the language of absence while quietly filling it with activity.

This linguistic compromise allowed continuity. The vacuum could remain "empty" in name while functioning as something rich and dynamic in practice. The contradiction was tolerated because it was productive. It allowed equations to work without demanding a philosophical reckoning.

But the cost of this compromise was conceptual clarity.

When emptiness behaves like substance, the distinction between what exists and what is described becomes blurred. The vacuum ceases to be a simple absence and becomes an unnamed participant in physical processes. Yet because it is still called "nothing," its role remains obscured.

This ambiguity has consequences. It limits the questions we are willing to ask. If the vacuum is assumed to be absence, then its influence must be treated as secondary or emergent. If it is allowed to be something, then its properties demand explanation. The choice between these interpretations shapes the direction of inquiry.

The inherited assumption of emptiness thus acts as a quiet constraint. It frames how problems are posed and which explanations are considered acceptable. It encourages us to treat the vacuum as a background rather than a component, even when evidence suggests otherwise.

Recognizing this assumption is the first step toward loosening its grip. It does not require rejecting established physics. It requires acknowledging that a concept once adopted for convenience may have outlived its usefulness.

The vacuum was called empty because it was easier to think that way. But ease is not evidence. As understanding deepens, the language that once served us may need to evolve.

This section does not argue that the vacuum is a substance in the traditional sense. It argues that treating it as absence has limited our questions. By recognizing that limitation, we create space—conceptual space—to ask what the vacuum might actually be, rather than what it is not.

8.2 — Consequences Without a Medium

Once the vacuum was defined as absence, physics faced an unavoidable dilemma: how do effects propagate through what has no substance? If nothing exists between objects, then nothing should transmit influence. And yet, influence was everywhere—forces acted, light traveled, energy moved, and changes propagated across vast distances.

To reconcile this contradiction, physics developed a subtle workaround. Instead of redefining the vacuum, it redefined what could happen *within* it.

This is where the idea of action without medium quietly took root.

Classical physics had relied on contact or medium-based transmission. Sound required air. Waves required water or string. Motion implied something that could carry disturbance. But once space was declared empty, the burden shifted. Effects had to propagate without a carrier. Fields emerged as the solution—a way to describe influence without substance.

Fields were brilliant inventions. They preserved continuity without reintroducing matter. They allowed interactions to occur across space without requiring a material medium. A field could exist everywhere without being "made of" anything. It could vary, store energy, and transmit effects while remaining conceptually weightless.

This move solved immediate problems, but it also introduced a subtle conceptual inversion.

Instead of asking what physically exists between objects, physics began to describe how objects behave *as if* something existed between them. The field became a mathematical intermediary—a tool for calculation rather than a declared entity. It carried influence without needing a physical carrier.

This worked remarkably well. So well, in fact, that the absence of a medium stopped being seen as a limitation and became an assumption.

But the cost of this move was deferred understanding.

By removing the need for a physical medium, physics freed itself from explaining how interactions propagate. The focus shifted to describing how they appear to behave. This was an effective strategy for prediction, but it left an unresolved question at the foundation: *what actually transmits influence across space?*

Rather than confronting this directly, the field absorbed the ambiguity. The vacuum became a placeholder—neither something nor nothing. It was allowed to possess properties without being considered a substance. It could fluctuate, store energy, and affect particles, yet it remained conceptually empty.

This ambiguity allowed theory to advance rapidly. It also masked the absence of a physical mechanism.

As long as equations matched observations, the lack of an underlying medium was tolerated. Theoretical consistency took precedence over ontological clarity. Physics moved forward, but with a growing conceptual gap between mathematical description and physical intuition.

The consequence of this approach is subtle but profound. By treating the vacuum as a non-entity with properties, physics avoided confronting a deeper question: what is the carrier of physical influence? If energy propagates, something must change. If forces act, something must respond. If fields exist, they must exist *in* something—or be something themselves.

By declining to address this directly, modern theory gained flexibility but lost interpretive grounding.

This is not an accusation of error. It is an acknowledgment of a strategic choice. The choice allowed extraordinary progress. It also postponed fundamental clarity.

The concept of the vacuum became a placeholder—an abstract space where equations operated without requiring a physical substrate. Over time, this placeholder acquired more and more responsibility. It became the stage, the medium, and the participant, all without ever being defined as such.

This is the quiet contradiction at the heart of modern physics. The vacuum is treated as both nothing and everything, depending on what the equations demand. That dual role is not sustainable indefinitely.

Recognizing this does not undermine the achievements built upon it. It simply reveals the next question waiting to be asked: if the vacuum is not nothing, then what is it?

This question does not negate existing theories. It completes them.

8.3 — Why Nothing Had to Do Everything

Once the vacuum was defined as absence, it inherited an impossible responsibility. It had to explain everything while remaining nothing. This contradiction did not arise from confusion, but from necessity. Physics needed a background that did not interfere, yet somehow allowed all interactions to occur. In giving the vacuum no substance, it was simultaneously given every role.

This is where conceptual overload began.

When space was stripped of materiality, it did not become irrelevant. Instead, it became the silent stage on which all physical processes played out. Fields propagated through it. Forces acted across it. Energy fluctuated within it. And yet, because it was defined as "nothing," these behaviors had no physical carrier. The vacuum became a paradox: an entity that could do everything while being nothing.

To manage this contradiction, physics adopted a subtle maneuver. Instead of treating the vacuum as an object, it treated it as a condition. The vacuum was not something that acted, but something in which action occurred. This allowed equations to function without assigning substance. The problem was deferred, not solved.

As theoretical demands grew, so did the responsibilities placed upon this conceptual emptiness. The vacuum became the reference state for energy, the medium for quantum fluctuations, the background against which curvature was defined, and the arena in which fields interacted. Each new role added explanatory power, but also increased conceptual strain.

Eventually, the vacuum was doing too much while being allowed to be nothing.

This is the heart of the issue. When a single concept is required to perform incompatible roles—absence and activity, neutrality and agency—it becomes overloaded. It can no longer be clearly defined, because any definition would contradict one of its functions. The result is a framework that works mathematically but resists intuitive understanding.

This overload did not occur because physicists were careless. It occurred because the alternative—redefining the vacuum as something real—would have required a foundational shift. That shift would have challenged long-standing assumptions about what constitutes physical existence. Rather than take that step, the discipline stretched the meaning of "nothing" to accommodate necessity.

Over time, this stretching became normalized. The vacuum was spoken of as energetic yet empty, active yet passive, structured yet featureless. These contradictions were tolerated because they were useful. They allowed theories to operate without demanding a rethinking of ontology.

But conceptual debt accumulates.

When a framework relies on a term that must simultaneously mean everything and nothing, it begins to lose explanatory precision. The vacuum becomes a placeholder for unresolved structure. It absorbs unanswered questions instead of answering them.

This does not mean the theory fails. It means the language has reached its limit.

At this point, the role of interpretation becomes critical. One can either continue layering abstractions atop an undefined foundation or pause to reconsider the foundation itself. The second option requires acknowledging that what has been treated as emptiness may, in fact, be a form of structure not yet named.

The idea that "nothing" does the work of "something" is not an error—it is a signpost. It indicates that the conceptual tools in use have reached the boundary of their effectiveness. Beyond that boundary lies the need for a new description, one that does not rely on absence to explain presence.

This section does not propose such a description yet. It simply clarifies the tension. The vacuum has been asked to do too much while being denied any identity. Recognizing this imbalance is the first step toward resolving it.

What comes next is not a rejection of established theory, but an invitation to reconsider what we have been calling "nothing."

8.4 — The Need for a Physical Substrate

Once the vacuum is recognized as doing more than merely *existing*, the question it raises becomes unavoidable: if something behaves as though it has structure, response, and continuity, on what basis do we continue to call it nothing?

This question does not arise from dissatisfaction with existing theory. It arises from fidelity to it. When a framework repeatedly attributes physical consequences to an entity, intellectual honesty requires that we reconsider what that entity represents. At some point, the language of absence becomes insufficient to carry the weight of function.

A physical substrate need not be a substance in the classical sense. It need not be particulate, tangible, or directly observable. What matters is not materiality, but *capacity*. A substrate, in this context, is anything capable of sustaining interaction, storing influence, or mediating change. The moment the vacuum is asked to do these things, it begins to behave as such a substrate—whether or not we choose to name it.

This is not a claim that the vacuum "is something" in a naive or mechanical way. It is a recognition that complete emptiness cannot host structure. If effects propagate, if energy redistributes, if correlations persist, then something must be enabling those processes—even if that "something" does not resemble matter as we traditionally understand it.

Physics already acknowledges this implicitly. Quantum field theory treats the vacuum as a state with structure, fluctuations, and measurable consequences. Cosmology treats it as a reservoir influencing expansion. Yet these roles are often treated as exceptions rather than indicators of a deeper ontology. The vacuum is allowed to act, but not to *be*.

This hesitation arises from an understandable caution. Historically, attempts to reify the medium of space—ether theories, for example—led to conceptual confusion and experimental contradiction. The lesson drawn was not merely that a particular model failed, but that speaking of a medium at all was misguided. As a result, physics adopted a posture of deliberate silence on the question of what underlies spacetime.

That silence, however, came with a cost. By refusing to speak of structure, physics limited itself to describing relationships without explaining why those relationships exist or behave as they do. The vacuum became an operational placeholder rather than an ontological participant.

The situation now invites reconsideration, not regression. The goal is not to resurrect old mechanical ether models or replace one metaphysical substance with another. It is to acknowledge that the role assigned to "nothing" has expanded beyond what emptiness can coherently support.

A physical substrate need not be material, granular, or localized. It need only possess the capacity to support interaction, store potential, and mediate change. In this sense, the idea of a structured vacuum is not an addition to physics but a clarification of what has already been implicitly assumed.

This shift does not demand new particles or forces. It demands a reclassification of what counts as physical. The vacuum need not be an object, but it may be an active state—a structured background whose properties give rise to the phenomena we observe.

Such a view aligns naturally with the way modern theories already behave. Fields are defined everywhere. Energy exists even in the absence of particles. Spacetime responds to conditions placed upon it. These are not features of emptiness. They are features of an underlying medium, even if that medium does not resemble anything tangible.

Recognizing this does not invalidate existing models. It reframes them. It allows the vacuum to be treated not as a mathematical trick, but as a physical context with its own characteristics. This reframing opens conceptual space for understanding why certain behaviors arise without requiring ad hoc additions.

Importantly, acknowledging a physical substrate does not commit one to a specific model. It simply restores coherence between what theories describe and what they implicitly assume. It bridges the gap between mathematical representation and physical intuition.

This is where the discussion naturally leads. Once the vacuum is allowed to be more than an absence, the question shifts from "what fills empty space?" to "what kind of structure makes such behavior possible?" That shift does not reject existing physics—it deepens it.

The next step is not to define the substrate prematurely, but to recognize that the absence of definition has itself become a limitation. The vacuum, long treated as nothing, now stands revealed as something that demands understanding.

What follows is not a departure from established science, but an invitation to extend it—carefully, critically, and without the burden of old assumptions.

8.5 — The Question Einstein Left for the Future

Einstein did not leave behind a finished picture of reality. He left behind a direction.

Despite the reverence that surrounds his work, Einstein himself was acutely aware of its incompleteness. He knew that his theories, powerful as they were, described *behavior* rather than *essence*. They explained how the universe behaves, not what fundamentally gives rise to that behavior. This distinction mattered deeply to him, even when it was uncomfortable to pursue.

Throughout his later years, Einstein searched for a unifying understanding that could connect gravitation, electromagnetism, and the deeper structure of space itself. This search was not driven by ambition or dissatisfaction with his own achievements, but by a recognition that something essential remained unarticulated. The equations worked. The predictions held. And yet, something fundamental was missing.

What he sensed—but could not formalize—was that spacetime geometry alone could not be the final explanation. Geometry described how phenomena unfolded, but not why they had the capacity to unfold at all. The equations mapped behavior, but the nature of the stage on which that behavior occurred remained undefined.

Einstein understood this limitation. He spoke often of his discomfort with purely formal explanations. He resisted interpretations that replaced understanding with calculation. For him, physics was not merely about prediction; it was about intelligibility. And intelligibility required more than mathematical success—it required conceptual grounding.

This is where his work ultimately stopped—not because he failed, but because he reached the edge of what his framework could support. He had described the curvature of spacetime with extraordinary precision, yet the question of *what* spacetime fundamentally is remained unanswered. The vacuum, the medium in which everything occurred, was still treated as a passive backdrop rather than an entity with its own nature.

That unresolved gap is not an oversight. It is a boundary.

Einstein left behind a theory that worked brilliantly within its domain and a question he did not attempt to close: *What is the physical nature of the medium whose geometry we describe?* He resisted speculative answers not because the question was unimportant, but because he recognized the danger of premature closure.

This restraint is often overlooked. Einstein is frequently portrayed as having provided a complete vision of spacetime. In reality, he left an open door—perhaps deliberately. His work clarified what could be said with rigor, and by doing so, illuminated what could not yet be said at all.

The question he left behind is not a failure of imagination but a marker of intellectual honesty. It acknowledges that description can outrun explanation, and that science progresses not only by solving problems but by recognizing when a deeper level of understanding is required.

This is where the present inquiry begins.

The question is no longer whether spacetime curves or how it behaves under energy and mass. Those insights are established. The question is what kind of entity must exist for such behavior to be possible at all.

This is not a departure from Einstein's legacy. It is a continuation of it.

The framework introduced in this work does not seek to replace relativity or diminish its achievements. It seeks to extend the conversation Einstein himself left open—by asking what kind of underlying structure could give rise to the phenomena he so elegantly described.

In this sense, the transition to the next chapter is not a break, but a handoff. A recognition that the language of geometry has taken us as far as it can, and that a deeper description—one concerned with the nature of the substrate itself—is now required.

What follows is not an answer imposed upon Einstein's work, but a response to the question he left behind.

Chapter 9 — Universal Energy Dynamics as Completion

"Every deep theory is a confession of what it could not yet explain."

Completion, Not Correction

Every major framework in science eventually reaches a moment of quiet tension—when it no longer fails, yet no longer fully satisfies. Not because it is wrong, but because it has said everything it can say in its own language.

This is the moment at which understanding does not collapse; it matures.

Universal Energy Dynamics does not arise as a rebellion against existing physics, nor as an attempt to overturn its foundations. It emerges from a more restrained realization: that the prevailing frameworks have reached the edge of what they were designed to explain. Their success has revealed their boundary.

UED is not offered as an alternative to relativity, quantum theory, or cosmology. It does not replace them, correct them, or compete with them. Instead, it steps into the conceptual space they leave open—the space between description and origin, between mathematical behavior and physical meaning.

Where earlier frameworks focused on *how* phenomena behave, UED asks *what must exist* for such behavior to occur at all.

This distinction matters. It reframes the entire discussion. Rather than treating the universe as a system of interactions happening within an undefined background, UED treats the background itself as a structured, dynamic participant. Not as a substance in the classical sense, but as an organized field of potential that gives rise to observable effects.

In this view, expansion, curvature, and energy are not primary actors. They are expressions—consequences of deeper organizational dynamics. The vacuum is no longer an empty stage, but a structured medium whose internal balance determines how phenomena appear and evolve.

This is not a break from science, but a continuation of its trajectory. Every major advance has come from recognizing that what was once treated as background was, in fact, fundamental. Space replaced ether. Fields replaced forces. Geometry replaced absolute reference frames. Each shift reframed the question rather than answering it outright.

Universal Energy Dynamics follows this same pattern.

It does not challenge established results; it reframes their context. It does not dispute measurements; it reinterprets what those measurements signify. It does not compete with existing theories; it completes the picture they began to draw.

In this sense, UED is not a new theory layered atop old ones. It is a lens—a way of understanding why the existing descriptions work as well as they do, and why they begin to strain at certain limits.

This chapter begins the transition from critique to construction. Having examined where existing frameworks reach their explanatory boundaries, we now explore what lies just beyond them—not as speculation, but as a natural continuation of the questions those frameworks themselves leave open.

What follows is not a departure from physics, but a return to its foundational intent: to describe reality as it is, not merely as it appears.

9.1 — From Forces to Equilibrium

For centuries, physics has been built around the language of force. Motion was explained as the result of pushes and pulls, interactions between entities that acted upon one another across space. This framework proved extraordinarily successful. From planetary motion to electromagnetism, the idea of forces acting between objects offered clarity, predictability, and mathematical elegance. It shaped not only equations, but intuition itself.

Yet embedded within this success was a subtle limitation—one that remained invisible precisely because the framework worked so well.

Force-based thinking assumes that change occurs because something *acts upon* something else. A body moves because a force acts on it. A field influences matter because it exerts influence. This worldview treats interaction as primary and treats the background as passive. It asks: *What is pushing what?* and *how strong is the push?*

But over time, the scope of physics expanded beyond the domain where such questions are sufficient. As attention shifted from objects to fields, and from fields to spacetime itself, the language of force began to show strain. What does it mean for spacetime to "act" on matter? What pushes curvature? What exerts force in a vacuum that contains nothing?

These questions did not expose errors in physics, but they revealed a mismatch between conceptual tools and the phenomena being described.

The transition toward equilibrium-based thinking begins at this point—not as a rejection of force, but as an expansion of perspective. Instead of asking what pushes what, the focus shifts to how systems organize themselves. Rather than seeing change as imposed from outside, change is understood as emerging from internal structure and balance.

In this view, motion is not the result of applied force alone, but the natural tendency of systems to adjust toward equilibrium configurations. Forces become expressions of imbalance, not fundamental drivers. They describe how systems respond when displaced from preferred states.

This shift subtly but profoundly alters interpretation.

Under a force-centric framework, the universe is a stage of interactions. Under an equilibrium-based framework, the universe is a network of conditions seeking balance. Motion, curvature, and energy flow are not imposed but arise from relational structure.

This is not a rejection of classical mechanics or field theory. It is a reorientation of emphasis. Forces remain valid descriptions of local behavior, but they are no longer the deepest explanatory layer. They become manifestations of deeper organizing tendencies within the system.

This shift becomes especially important when dealing with phenomena that do not fit neatly into force-based intuition—such as cosmic expansion, vacuum energy, or spacetime curvature. In these contexts, it becomes increasingly difficult to identify a "force" in the traditional sense. There is no clear agent pushing space apart, no identifiable source exerting pressure. Yet change occurs.

An equilibrium-based perspective reframes this puzzle. Instead of asking what force causes expansion, it asks what configuration the system is moving toward. Expansion becomes not an action but a response—a relaxation toward a state of lower tension or higher symmetry within a structured medium.

This perspective aligns naturally with the idea that the universe is not a collection of isolated interactions, but an integrated system whose behavior emerges from internal coherence. The emphasis shifts from dynamics imposed from without to dynamics arising from within.

Such a shift does not eliminate forces; it contextualizes them. Forces become localized expressions of a deeper tendency toward balance. They are how equilibrium is restored when disturbed, not the origin of motion itself.

This way of thinking also dissolves certain conceptual paradoxes. If the vacuum is understood as an active medium rather than empty space, then its influence on matter is no longer mysterious. Energy does not need to be "carried" across emptiness; it propagates through a structured background that supports its movement.

In this light, the transition from force-based explanations to equilibrium-based ones is not a rejection of physics but its maturation. It reflects an increasing recognition that the universe behaves less like a machine with moving parts and more like a self-regulating system seeking coherence.

Universal Energy Dynamics builds on this realization. It does not deny forces, but it reframes them as expressions of underlying equilibrium processes. Forces become symptoms rather than causes—visible indicators of deeper structural adjustments within a continuous field.

This reframing allows long-standing puzzles to be approached from a new angle. Instead of asking why forces act at a distance, one asks how equilibrium is maintained across a connected structure. Instead of invoking external causes, one looks for internal consistency.

In doing so, the focus shifts from motion to meaning—from describing how things move to understanding why they must move the way they do.

This marks a significant conceptual transition. It does not discard the achievements of physics; it situates them within a broader explanatory context. Forces remain indispensable, but they are no longer the final story.

The universe, viewed through this lens, is not a collection of interacting parts but a coherent system striving for balance. Motion becomes the language of that striving. And equilibrium, rather than being static or inert, becomes the dynamic principle underlying all change.

This shift—from force to equilibrium—is not a rejection of the past. It is the next step in understanding what the past has already revealed.

9.2 — Structured Vacuum Energy

Once the idea of equilibrium replaces force as the primary lens, the next question arises naturally: *what is it that seeks equilibrium?* If motion, curvature, and interaction are not imposed from outside but emerge from internal balance, then there must exist something that can hold imbalance, respond to it, and restore coherence. That "something" is what we refer to as **Structured Vacuum Energy**.

This term does not describe a new particle, substance, or exotic field. It describes a condition—a state of the underlying fabric of reality that is capable of organization, response, and continuity. The vacuum, in this view, is not an empty stage but a structured medium whose internal state governs how energy manifests and evolves.

Traditional physics has long acknowledged that the vacuum is not truly empty. Quantum fluctuations, zero-point energy, and field activity all point toward an underlying dynamism. Yet these features are often treated as anomalies—side effects of equations rather than primary attributes of reality. Structured Vacuum Energy reframes this relationship. It proposes that what has been treated as background noise is, in fact, the foundational medium itself.

To understand this, it helps to move away from thinking of the vacuum as a void and toward thinking of it as a *state*. Just as temperature describes the collective behavior of particles, Structured Vacuum Energy describes the condition of the underlying field that gives rise to observable phenomena. It is not a substance moving through space, but a configuration of space itself.

In this view, space is not empty; it is organized. It possesses internal degrees of freedom that determine how energy propagates, how fields interact, and how matter stabilizes. These degrees of freedom are not particles or forces but structural tendencies—patterns of alignment and response that give rise to physical behavior.

The word "structured" is essential here. It implies order without requiring rigidity. The vacuum is not uniform in the simplistic sense, nor is it chaotic. It is dynamically balanced. Its structure allows for variation while maintaining coherence. This balance is what permits both stability and change.

When energy moves through such a medium, it does not do so arbitrarily. It interacts with the structure, adjusting and being adjusted in return. What we observe as propagation, curvature, or expansion can be understood as the system's response to internal stress or imbalance. Motion, in this framework, is not imposed—it emerges.

This perspective also reframes the concept of energy itself. Energy is no longer something that travels through space like a substance in a container. Instead, it represents a localized state of excitation within the structured vacuum. Movement is not the transport of energy across emptiness but the reconfiguration of the underlying medium.

Such a view naturally accommodates phenomena that otherwise appear paradoxical. Non-local correlations, field interactions without apparent carriers, and the persistence of structure across vast distances all become intelligible when the vacuum is understood as an interconnected system rather than an empty void.

Importantly, this does not contradict existing theories. Quantum field theory already treats particles as excitations of underlying fields. General relativity describes spacetime as responsive to energy and momentum. Structured Vacuum Energy simply unifies these insights under a single conceptual umbrella. It states explicitly what is often implicit: that the vacuum is not a passive background but an active participant.

This perspective also clarifies why attempts to isolate "pure emptiness" have consistently failed. Every effort to probe deeper reveals structure, fluctuation, or constraint. The absence of matter does not lead to the absence of physics. Instead, it reveals the most fundamental level at which physics operates.

By acknowledging the vacuum as structured, one can move beyond the false dichotomy between matter and space. They become expressions of the same underlying reality, differentiated by state rather than substance. This shift dissolves long-standing conceptual tensions without discarding empirical success.

In this framework, the universe is not built upon emptiness punctuated by objects. It is a continuous field of organized potential, within which matter, energy, and motion arise as expressions of configuration rather than entities imposed from without.

Structured Vacuum Energy thus serves as a unifying concept. It bridges the gap between fields and geometry, between quantum behavior and cosmic structure. It offers a way to speak about the deep substrate of reality without resorting to metaphysical speculation or abandoning empirical discipline.

This is not an assertion of final truth, but a reorientation of perspective. It invites us to stop asking what fills the vacuum and instead ask how the vacuum itself is organized. In doing so, it opens a path toward understanding the universe not as a collection of things moving through nothing, but as a coherent system whose behavior emerges from its own internal structure.

That shift—from emptiness to structure—is the quiet but profound step that underlies the concept of Structured Vacuum Energy.

9.3 — Gravity Revisited Without Geometry Worship

Few ideas in science have been as successful—and as quietly misunderstood—as Einstein's description of gravity. General Relativity reshaped our understanding of motion, space, and time with unmatched elegance. Its predictions have been verified repeatedly, from the bending of light to the behavior of gravitational waves. And yet, the success of the theory has also encouraged a subtle misunderstanding: that geometry itself *is* gravity.

This chapter does not challenge the accuracy of General Relativity. It challenges the interpretation we have attached to it.

Einstein showed that gravity does not behave like a conventional force. Objects do not attract one another across space in the Newtonian sense; instead, they follow paths determined by the curvature of spacetime. This insight was revolutionary. It replaced force with geometry, action with structure, and interaction with configuration. The equations worked—and they still do.

But success bred a quiet assumption: that geometry is not just a description of gravitational behavior, but its cause.

This is where the conceptual leap occurred. What began as a powerful mathematical representation gradually hardened into an ontological claim. Curvature ceased to be a way of describing gravitational behavior and became the thing that causes it. Geometry was no longer a language; it was treated as an actor.

This shift was subtle, almost invisible, because it produced correct results. Yet it introduced a quiet confusion between *description* and *agency*.

In General Relativity, spacetime geometry encodes how matter and energy influence motion. The equations tell us how trajectories respond to curvature. But they do not explain why curvature exists in the first place. Geometry describes relationships; it does not generate them. When we say that mass "curves spacetime," we are using a metaphor that compresses cause and effect into a single phrase.

The theory works because it captures relational structure, not because it identifies an underlying mechanism. This distinction is often overlooked because the mathematics is so effective. But effectiveness should not be mistaken for completeness.

When geometry is treated as the fundamental agent, deeper questions are quietly deferred. What gives spacetime its ability to curve? Why does energy affect geometry at all? What medium supports this geometric behavior? These questions are not answered by the equations; they are bracketed out.

This is not a flaw of General Relativity. It is a boundary condition of its design.

Einstein himself understood this. He never claimed that spacetime geometry was the ultimate substrate of reality. His later work reflects a persistent dissatisfaction with purely geometric explanations. He sought a deeper unifying structure, one that could account for both geometry and matter within a single framework.

The problem arises not from Einstein's theory, but from how it has been interpreted since. Over time, the mathematical success of the equations encouraged a form of conceptual complacency. Geometry became reified. The language of curvature hardened into an ontology.

This is what might be called "geometry worship"—the tendency to treat mathematical description as ontological explanation. It is not an error of calculation, but of interpretation. Geometry tells us how things behave relative to one another, not why such relations exist in the first place.

When gravity is described purely as geometry, it appears to act without a medium. Space curves itself, somehow, without an underlying structure. This works mathematically, but it leaves an explanatory vacuum. The deeper question—*what allows geometry to have physical effect*—is left unanswered.

Reintroducing a structured substrate does not undermine relativity; it clarifies it. Geometry can then be understood as the manifestation of deeper organizational principles rather than their cause. The curvature of spacetime becomes a macroscopic expression of underlying equilibrium dynamics, not an inexplicable property of emptiness.

In this view, General Relativity remains entirely valid within its domain. Its equations describe how geometry responds to energy and momentum. What changes is the interpretation of what that geometry represents. It becomes a descriptive layer rather than the foundational one.

This distinction allows us to preserve everything that works in relativistic physics while opening conceptual space for deeper explanation. It acknowledges the power of Einstein's framework without elevating it to metaphysical finality.

By releasing the idea that geometry must be fundamental, we gain the freedom to ask what underlies it. We stop asking how spacetime curves and begin asking why curvature is the form that response takes.

That shift—from geometry as cause to geometry as expression—is the key to moving forward. It allows gravity to be understood not as a mysterious force embedded in space, but as the visible outcome of deeper structural dynamics.

In this way, gravity is not diminished. It is contextualized. It becomes one layer in a hierarchy of description, not the ultimate ground of reality.

This perspective does not overturn Einstein's work. It completes it by placing it where it belongs: as a powerful description of behavior, not the final word on what existence is made of.

And with that reframing, the door opens to a deeper understanding of the universe—one in which geometry is not worshipped, but understood as a consequence of something more fundamental.

9.4 — A Still Universe with Dynamic Consequences

The idea that the universe might be fundamentally *still* feels counterintuitive, even heretical, in a scientific culture built on motion. From the earliest observations of the heavens, change has been equated with movement—stars drifting, galaxies receding, space expanding. Motion became synonymous with reality itself. To suggest that the universe could be fundamentally still seems, at first glance, to contradict everything we observe.

Yet this discomfort arises from a quiet assumption: that motion must always originate from movement of substance through space. What if this assumption is incomplete?

A still universe does not imply a static or frozen cosmos. It implies something more subtle—that change arises not from objects moving through an empty backdrop, but from shifts within an underlying structured state. In such a view, motion is not the displacement of things through space but the reconfiguration of relationships within a stable framework.

This distinction is critical.

If the universe is treated as a dynamic field rather than a container of moving parts, then what we interpret as motion may instead be the visible expression of internal rebalancing. Expansion, curvature, and flow become manifestations of how the system redistributes itself, not evidence of space itself stretching or moving.

In this framework, stillness does not mean inactivity. It means coherence. A still lake can support waves without itself traveling. The water molecules oscillate, but the body of water remains in place. Similarly, a structured vacuum can support dynamic processes—energy propagation, particle interactions, curvature—without itself undergoing translational motion.

This analogy helps clarify a crucial point: motion does not require displacement of the medium as a whole. It requires only that the medium can support variation. The universe, in this view, is not expanding like a balloon but expressing changes within a stable underlying structure.

Such a perspective reframes many long-standing interpretations. Redshift, for example, need not be interpreted as evidence of space itself stretching. It can be seen as the cumulative effect of energy interacting with a structured medium over vast distances. The apparent expansion becomes a measure of energy redistribution rather than geometric dilation.

This does not deny observed phenomena; it reframes their cause. Observational data remain intact. What changes is the story we tell about what those observations mean.

In a still-universe framework, motion is local, not global. Galaxies move relative to one another, energy flows, structures evolve—but the underlying fabric remains stable. The universe is dynamic without being in flux. It changes without "going somewhere."

This distinction matters because it resolves several conceptual tensions. It removes the need for space itself to possess agency.

It avoids treating expansion as a force-like entity. And it allows conservation principles to be applied more naturally, without invoking compensatory mechanisms.

Most importantly, it restores coherence between micro- and macro-scale descriptions. If the same structured medium underlies all interactions, then the behaviors observed at different scales are expressions of the same foundational dynamics, not separate regimes stitched together by convenience.

A still universe is not a frozen universe. It is a dynamically balanced one. Its activity arises from internal relationships rather than external motion. Change is real, but it is relational rather than translational.

This perspective also aligns naturally with the idea of equilibrium discussed earlier. Equilibrium does not mean stasis; it means regulated change. A system at equilibrium can be rich with motion, provided that the motion does not accumulate into unbalanced drift.

In this sense, the universe can be both active and stable, dynamic yet grounded. The apparent paradox dissolves once motion is understood as internal reconfiguration rather than displacement through emptiness.

Reframing the cosmos in this way allows long-standing puzzles to be approached differently. Expansion becomes a manifestation of structured response. Curvature becomes a pattern of equilibrium. Energy becomes a mode of organization rather than a thing in transit.

This is not a rejection of existing cosmology but a reinterpretation of its meaning. It suggests that what we have been observing all along may not be a universe flying apart, but a universe continuously rebalancing itself within a deeper, steady framework.

In this light, stillness is not the absence of motion—it is the condition that makes motion intelligible.

9.5 — Why This Completes the Sentence

Every scientific framework tells a story. Not a story in the literary sense, but a narrative of understanding—where phenomena come from, how they behave, and why they appear as they do. Some stories end with a period. Others end with a comma, inviting continuation. The evolution of physics has largely been a sequence of such continuations, each theory extending the previous one without truly closing the sentence.

Universal Energy Dynamics does not arrive to replace what came before. It arrives to complete the sentence that earlier frameworks were never meant to finish.

The success of modern physics lies in its ability to describe behavior with extraordinary precision. Classical mechanics described motion. Relativity described structure. Quantum theory described interaction. Each framework clarified a layer of reality while leaving deeper questions intact. None of them claimed to explain *why* the universe is capable of behaving at all. They explained *how* behavior unfolds once it exists.

UED enters precisely at this boundary—not to contradict, but to contextualize.

Where previous theories describe motion, UED addresses the condition that makes motion meaningful. Where equations define relationships, UED addresses what allows relationships to exist in the first place. It does not compete with established models; it gives them a shared foundation.

This distinction is crucial. Replacement implies error. Completion implies continuity.

The role of UED is not to rewrite equations or dismiss established results. It is to interpret them within a broader conceptual frame. It answers the question that classical and modern physics quietly set aside: *What must reality be like for these laws to hold at all?*

In this sense, UED does not challenge Einstein's insights—it fulfills them. Einstein showed that gravity is not a force but a manifestation of structure. UED extends this logic by asking what makes structure possible in the first place. It moves one layer deeper, from geometry to the conditions that allow geometry to arise.

This move resolves tensions without erasing prior knowledge. Expansion remains observable. Curvature remains measurable. Energy remains conserved. What changes is how these features are interpreted—not as independent actors, but as expressions of a deeper organizational principle.

By framing the universe as a structured equilibrium rather than a collection of forces acting in empty space, UED provides coherence where fragmentation once existed. It explains why different domains of physics can describe reality so effectively while seeming conceptually disconnected. They are describing different aspects of the same underlying order.

This is why UED does not seek to overthrow existing theories. It seeks to give them a shared ground. It treats them as approximations valid within their domains, each revealing a facet of a more general structure.

In this sense, UED functions not as a rival framework, but as a unifying perspective. It allows diverse models to coexist without contradiction by placing them within a broader conceptual context. It turns apparent conflicts into complementary descriptions.

Completion, here, does not mean finality. It means coherence. It means recognizing that the questions left open by earlier theories were not mistakes but invitations. Invitations to look beneath equations, beyond formalism, and toward the conditions that make both possible.

By framing the universe as a structured, dynamic equilibrium, UED offers a way to understand motion without invoking arbitrary forces, and change without invoking chaos. It restores continuity between description and meaning.

In doing so, it does not close inquiry—it stabilizes it. It provides a foundation from which deeper questions can be asked without contradiction or conceptual strain.

That is why UED is not a replacement, but a completion. Not an alternative path, but the next logical step in understanding what physics has been describing all along.

Interlude IV — From Structure to Meaning

There comes a moment in every serious inquiry when structure alone is no longer enough. The framework may be elegant, internally consistent, and capable of explaining vast domains of experience—yet something essential remains unspoken. At that point, understanding asks to move beyond structure and toward meaning.

Structure tells us *how* things are arranged. Meaning asks *why* that arrangement matters.

Throughout this work, we have moved carefully through layers of description—geometry, fields, equilibrium, dynamics—each offering clarity, each expanding our grasp of how the universe behaves. But structure, however refined, is only a map. It shows relationships, not reasons. It describes order, but not intention. It explains behavior, but not significance.

This is not a flaw. It is the natural limit of structural thinking.

Meaning emerges when structure is placed in context—when patterns are not only observed, but interpreted. Where structure answers *what happens*, meaning asks *what this says about the nature of reality itself.* The shift is subtle but profound. It marks the transition from analysis to understanding.

At this point in the journey, the foundational architecture has been laid. The reader has seen how description replaces explanation, how equilibrium replaces force, how structure replaces substance. The next step is not to add more structure, but to reflect on what this structure implies.

Meaning does not arise from complexity. It arises from coherence. When disparate ideas align—when descriptions across different domains begin to echo one another—a deeper understanding becomes possible. This is the threshold we now approach.

This is also the point where reductionism gives way to synthesis. The question is no longer how many parts compose the system, but how those parts relate in a way that makes the whole intelligible. The emphasis shifts from mechanics to significance.

Importantly, this is not a retreat into abstraction. It is a return to purpose. The goal is not to explain away mystery, but to locate it properly—to see where explanation ends and insight begins.

This interlude serves as a pause before that transition. It invites the reader to step back from analysis and prepare for integration. What follows is not a new set of arguments, but a reframing of what has already been established.

The movement from structure to meaning is not a leap; it is a natural continuation. It is what happens when understanding matures beyond description and begins to ask what all of this *means* for how we see the universe—and ourselves within it.

With this shift, the discussion moves from mechanisms to significance, from structure to sense. The journey does not end here. It deepens.

Part V — Closing The Circle

Returning to Einstein with clarity rather than reverence.

Chapter 10 — What Einstein Did Exactly

"A theory can be perfect in form and still incomplete in meaning." — Attributed to the spirit of Einstein's reflections

What Einstein Actually Did

By the time Albert Einstein completed his major work, the universe had not been explained—it had been reframed.

This distinction matters more than history often admits.

Einstein did not tell us what the universe *is*. He showed us how it behaves when observed through a particular lens. His genius lay not in declaring ultimate truths, but in redefining the rules by which physical descriptions could be made consistent. He replaced force with geometry, replaced absolute space with relational structure, and replaced intuition with mathematical coherence. What he offered was not a final picture of reality, but a new grammar for describing it.

And yet, over time, that grammar began to be mistaken for the language itself.

Einstein never claimed to have revealed the ultimate nature of existence. He did not say what spacetime *is*—only how it behaves when influenced by energy and matter. He did not define what causes curvature; he described how curvature manifests. He did not explain why the universe has the structure it does; he described the rules it appears to follow.

This distinction is subtle, but decisive.

The success of General Relativity was so complete that its descriptive power was mistaken for ontological closure. The equations worked so well that they became equated with reality itself. Geometry became substance. Mathematical consistency became physical explanation. Over time, interpretation hardened into assumption.

This chapter exists to gently unwind that conflation.

Einstein's achievement was not that he solved the universe, but that he showed how deeply structure governs behavior. He revealed that motion, gravity, and time itself could be understood as consequences of relational geometry. That insight changed physics forever. But it was never meant to be the final word.

What Einstein did was open a door, not close a case.

He provided a framework capable of extraordinary prediction, yet he remained cautious about its philosophical reach. He understood that equations describe patterns, not causes. He resisted claims that geometry itself was the ultimate reality. In his later years, he searched for a deeper unity—not because relativity failed, but because it succeeded too well while leaving deeper questions unanswered.

This chapter does not reinterpret Einstein's work as incomplete or flawed. It repositions it as a foundation rather than a destination. The achievements of relativity remain intact; what changes is how we understand their role.

The story that follows is not about correcting Einstein. It is about completing the arc he began—by recognizing the limits he himself acknowledged, and by extending his insights into a broader conceptual landscape.

This is not revisionism. It is continuation.

To understand what Einstein truly did, one must see both the brilliance of his constructions and the space he deliberately left open. That space—the space between description and essence—is where the next chapter of understanding begins.

10.1 — Einstein's True Achievement

To understand what Einstein truly achieved, one must first set aside both reverence and reduction. His work has been mythologized on one side and oversimplified on the other. In popular imagination, Einstein "explained the universe." In academic shorthand, he "replaced Newton." Both views miss the deeper truth. Einstein's real achievement was neither the overthrow of classical physics nor the final explanation of reality—it was the redefinition of how physical explanation itself should proceed.

At its core, Einstein's contribution was not the discovery of new forces or substances, but the recognition that *the structure of physical law matters more than the objects it describes*. He did not ask, "What is gravity made of?" He asked, "How must the world be structured for gravity to appear as it does?" This shift—from substance to structure—was revolutionary.

Before Einstein, physics treated space and time as passive backgrounds. Objects moved *within* them; forces acted *through* them. Even Newton, who revolutionized motion and mechanics, still assumed an absolute framework in which events unfolded. Einstein overturned this assumption. He showed that space and time are not fixed stages but participants in physical phenomena.

They bend, stretch, and relate differently depending on energy and motion.

This insight did not come from speculation but from a demand for internal consistency. Einstein noticed that existing laws contradicted one another when applied universally. Rather than modifying the laws piecemeal, he reexamined the conceptual foundations beneath them. His brilliance lay not in proposing exotic entities, but in asking what the equations were quietly implying.

The result was a framework where gravity was no longer a force acting across space, but a manifestation of spacetime geometry itself. Objects did not move because they were pulled; they moved because the structure of spacetime guided their trajectories. This was a profound shift. It replaced cause-and-effect narratives with relational ones.

Yet even here, Einstein was careful. He did not claim that geometry was the ultimate substance of reality. He used geometry as a language—a powerful one—to express observed regularities. His equations described how spacetime behaves in the presence of energy and momentum. They did not explain why spacetime exists, nor why it has the properties it does.

This distinction is crucial. Einstein provided a map, not the terrain. The map was extraordinarily accurate, but it did not claim to be the territory itself.

Over time, however, the clarity of this distinction blurred. The success of General Relativity encouraged the belief that geometry *is* reality, rather than a representation of it. The model became reified. What was once a descriptive framework began to be treated as an ontological statement about what the universe fundamentally is.

This was never Einstein's intention.

His later work reveals a man deeply aware of the incompleteness of his theory. He spent decades searching for a deeper unification—not because relativity failed, but because he believed that a deeper layer of explanation existed beneath it. He was not satisfied with describing how spacetime curves; he wanted to understand what made curvature possible at all.

Einstein's discomfort was philosophical as much as scientific. He sensed that equations alone could not fully account for physical reality. They captured patterns, but not origins. They described relations, but not the nature of what was related.

In this sense, his work represents both a pinnacle and a boundary. It marked the highest achievement of a certain mode of thinking while simultaneously revealing its limits. That is why his legacy is so enduring: it contains within it the seeds of its own extension.

Understanding Einstein properly means recognizing both his achievement and his restraint. He did not claim to have found the ultimate constituents of reality. He provided a framework that worked astonishingly well, and then stopped where certainty ended. That restraint is often overlooked, but it is central to his intellectual integrity.

What follows in the broader arc of understanding is not a rejection of Einstein's work, but a continuation of his method. To take his insights seriously is to recognize where they lead—and where they deliberately stop.

Einstein showed us how to describe the universe with unparalleled precision. The next question—one he implicitly left open—is what kind of reality makes such description possible in the first place.

That question marks the transition from explanation to understanding. It is where the story moves next.

10.2 — What He Did Not Claim

To understand Einstein clearly, one must pay attention not only to what he asserted, but also to what he carefully refused to assert. Much of the confusion surrounding his legacy arises not from his ideas themselves, but from the assumptions later generations attached to them. In many ways, Einstein has been over-interpreted—his equations turned into metaphysical statements he never intended to make.

Einstein did not claim to have discovered the ultimate nature of reality. He did not assert that spacetime *is* reality in its final form. He did not propose that geometry replaces substance, nor that mathematics exhausts meaning. What he offered was a framework—a way of describing how phenomena relate under certain conditions. The rest was inference layered upon inference.

One of the most persistent misreadings is the belief that Einstein equated geometry with physical reality itself. In truth, he treated geometry as a descriptive language, not as an ontological declaration. Spacetime curvature, in his work, was a mathematical representation of how objects behave under gravity, not a claim about what the universe fundamentally *is*. The distinction is subtle but essential.

Einstein was deeply cautious about metaphysical claims. He resisted any attempt to turn his equations into metaphysical dogma. When others suggested that spacetime itself *is* reality, or that matter is merely an illusion of geometry, he remained unconvinced. For him, theories were instruments—tools for organizing experience—not final truths about existence.

This caution is often forgotten because the elegance of general relativity invites philosophical interpretation. Its mathematical beauty tempts one to see it as a complete picture. But Einstein himself never made that leap. He understood that mathematics could describe relationships without revealing their ultimate cause.

Another common misunderstanding is the belief that Einstein replaced classical causality with something mystical or abstract. In fact, he was deeply uncomfortable with interpretations that removed causality or realism. He resisted probabilistic interpretations of quantum mechanics precisely because he believed that physical theories should describe an underlying reality, not merely predict outcomes.

This resistance reveals something essential: Einstein did not equate explanatory success with ontological truth. A theory could work exceptionally well and still be incomplete. For him, usefulness was not the same as finality.

Equally important is what Einstein did not claim about the vacuum. Although his equations allow spacetime to curve and respond to energy, he never claimed that spacetime itself was a substance with agency. He avoided language that would suggest the vacuum was an active entity. Instead, he treated spacetime geometry as a relational framework—a way of encoding how matter and energy interact.

This restraint is often lost in later interpretations that speak of spacetime as if it were a thing that stretches, flows, or bends of its own accord. Such language is metaphorical, not literal. Einstein understood this distinction, even when others blurred it for conceptual convenience.

His correspondence reveals a consistent unease with reifying mathematical constructs. He worried that physics might mistake its tools for reality itself. This concern guided his lifelong search for a deeper, unified description—one that would ground observed phenomena in something more fundamental than equations.

It is also important to note what Einstein did not claim about completeness. He never believed that general relativity was the final theory. On the contrary, he spent the latter part of his life searching for a more comprehensive framework that could unify gravity with other interactions. That search was motivated by a sense that something essential was still missing.

In this light, Einstein's work should be seen not as a finished structure, but as a boundary marker. He mapped the terrain with extraordinary precision, but he did not claim to have reached its end. His reluctance to declare victory is itself instructive.

The danger arises when later interpretations freeze his ideas into dogma. When mathematical elegance is mistaken for ontological finality, inquiry becomes constrained. The very theory that once expanded our understanding becomes a boundary that limits further questioning.

Einstein would have resisted such closure. His intellectual character was defined by skepticism—especially toward his own ideas. He valued questions over conclusions, coherence over comfort. To treat his work as the final word is to misunderstand his deepest commitment.

Recognizing what Einstein did *not* claim is therefore essential. It clears space for further inquiry. It allows us to respect his achievements without idolizing them. And it opens the possibility of extending his insights rather than merely repeating them.

In this sense, honoring Einstein means continuing the journey he began, not freezing it at a particular milestone. The next step does not reject his work—it builds upon the foundation he laid, guided by the same spirit of careful questioning that defined his own pursuit of understanding.

10.3 — Why His Work Endures

The endurance of Einstein's work is not a historical accident, nor is it the result of intellectual reverence alone. It persists because it was built on principles that remain resilient under scrutiny. Long after many theories have faded, his continue to function—not because they were perfect, but because they were fundamentally honest about what they could and could not explain.

At the heart of this durability lies a rare balance: Einstein's theories were precise without being rigid. They provided clear mathematical structure while leaving room for interpretation, refinement, and growth. This balance allowed them to remain relevant even as new evidence emerged and new conceptual frameworks developed.

One reason his work endures is that it was grounded in invariance. Rather than describing isolated phenomena, Einstein identified relationships that remained stable across frames of reference. These invariances—such as the constancy of physical laws under transformation—gave his theories a robustness that transcended specific observations. When later experiments confirmed predictions far beyond the conditions under which the theories were conceived, it became clear that the underlying principles were deeply rooted.

Another reason lies in the way his theories unify rather than fragment understanding. General relativity, for example, did not introduce a new force; it reframed gravity as a manifestation of geometry. This unification reduced complexity rather than adding to it. By showing that seemingly different phenomena could be understood through a single structural principle, Einstein's work gained a coherence that continues to withstand scrutiny.

Equally important is the openness of his framework. Einstein's theories were not closed systems that resisted reinterpretation. They were constructed with internal consistency but without metaphysical finality. This openness allowed later scientists to build upon them without contradiction. Quantum theory, cosmology, and modern astrophysics all interact with Einstein's ideas, sometimes in tension, but rarely in opposition.

His work endures because it does not depend on fragile assumptions. It does not rely on arbitrary parameters or unexplained constants inserted to force agreement with observation. Instead, it builds from symmetry, consistency, and empirical grounding. Even where adjustments have been necessary, the core structure has remained intact.

Another key factor in its longevity is humility. Einstein did not claim ultimate answers. He acknowledged uncertainty and left open questions deliberately unresolved. This humility made his work resilient. It invited extension rather than closure. When later discoveries challenged aspects of his framework, they did not undermine it; they enriched it.

The durability of his ideas also stems from their philosophical depth. They are not merely computational tools but conceptual frameworks that shape how we think about reality. They redefine notions of space, time, causality, and observation. Such foundational ideas do not expire quickly because they address how understanding itself is structured.

Perhaps most importantly, Einstein's work endures because it aligns with the way nature seems to behave. It does not force nature into preconceived forms; it listens to what observations imply and follows their logic rigorously. This fidelity to evidence, coupled with conceptual courage, gives his work a lasting quality that transcends technological change.

In this sense, Einstein's legacy is not a set of equations but a way of thinking. His theories endure because they are open frameworks rather than closed doctrines. They invite refinement rather than resist it. They provide clarity without claiming finality.

That is why, even as new ideas emerge and new paradigms are proposed, Einstein's work continues to serve as a foundation. Not because it answered every question, but because it asked the right ones—and left space for those questions to evolve.

In the end, the enduring power of Einstein's work lies not in what it solved, but in how it taught us to seek understanding. It remains relevant because it was never meant to be the last word, only a truthful one.

10.4 — From Consequences to Understanding

Every scientific revolution begins not with answers, but with consequences. Long before understanding arrives, effects announce themselves—patterns appear, regularities repeat, predictions hold. Only later does meaning catch up. This sequence is not a flaw in science; it is its natural rhythm.

Human inquiry almost always begins with observation, not explanation. We see motion before we know what causes it. We measure change long before we understand what changes. Science advances by first describing *what happens*, then slowly learning *why it happens*. The history of physics is, at its core, the gradual transformation of consequences into understanding.

This is the arc that Einstein's work followed—and the arc that modern physics now finds itself continuing.

Einstein's theories were not built from metaphysical speculation but from observable inconsistencies. He noticed that existing frameworks produced contradictions when pushed to their limits. Rather than inventing new entities, he reexamined the assumptions that framed those observations. His genius lay in recognizing that description precedes explanation. He did not claim to reveal the ultimate nature of reality; he refined how reality could be described without contradiction.

In doing so, he showed that consequences—such as time dilation, length contraction, and gravitational curvature—were not anomalies to be explained away but signals pointing toward deeper structure. These phenomena were not problems to solve but clues to interpret.

The modern challenge arises when such consequences are treated as endpoints rather than indicators. When predictions succeed repeatedly, it becomes tempting to treat them as explanations in themselves. The distinction between "what happens" and "why it happens" begins to blur. The description starts to masquerade as cause.

This is where scientific understanding risks stagnation—not because it fails, but because it succeeds too well.

The expansion of the universe, for example, is a precisely measured phenomenon. Its effects are observed across cosmic distances. Yet the explanation of why this expansion occurs, or what exactly is expanding, remains conceptually unresolved. The description functions flawlessly, but the interpretation remains provisional.

Such moments mark a transition point in scientific development. They indicate that a framework has reached the limit of what it can explain using its own assumptions. Progress then requires a shift in perspective rather than refinement of technique.

Historically, this is how science advances. Newtonian mechanics worked extraordinarily well until its assumptions were stretched beyond their domain. Classical thermodynamics described heat effectively before the microscopic nature of matter was understood. Each time, explanation lagged behind description until a deeper conceptual structure emerged.

We now find ourselves at a similar juncture.

The accumulated success of modern physics has created a landscape rich in data but increasingly sparse in interpretation. Theories describe behavior with astonishing precision, yet questions about underlying causes persist. This does not reflect failure; it reflects maturity.

Understanding emerges when we recognize that explanations evolve. What once served as a final answer becomes a stepping stone. Theories do not fail—they complete their role.

The movement from consequence to understanding requires a shift in perspective. Instead of asking how to force observations into existing frameworks, we ask what those observations are telling us about the limitations of our frameworks. This shift is subtle but profound. It transforms anomalies from problems into guides.

In this light, the persistence of unresolved questions is not a weakness of science but a sign of its vitality. It means that reality continues to challenge our models, inviting deeper insight.

The transition now underway is not about discarding established theories but about reframing their role. They remain accurate descriptions within their domains, yet they no longer claim to be the final word. Their success becomes the very reason to look beyond them.

This is the natural evolution of understanding: from measurement to meaning, from description to comprehension. Each stage prepares the ground for the next.

In this sense, the movement from consequence to understanding is not a departure from science—it is its fulfillment. It reflects the discipline's capacity to grow beyond its own achievements, to question its own foundations, and to refine its view of reality without abandoning rigor.

The journey continues not because science has failed, but because it has succeeded well enough to reveal what it does not yet explain. And in that recognition lies the path forward.

10.5 — An Open Ending, Not a Final Word

Every genuine inquiry ends not with certainty, but with orientation. When understanding deepens, it rarely closes a subject—it reframes it. This is where the journey now stands. Not at a conclusion, but at a threshold.

The ideas explored here do not seek to replace existing knowledge or declare a final truth. They seek to clarify what has been quietly unfolding beneath the surface of established theories. They ask not *what must be believed*, but *what must be re-examined.* In doing so, they restore a sense of openness that has always been essential to genuine inquiry.

The temptation to conclude—to finalize, to declare completeness—is strong. It offers comfort. It gives closure. But history shows that science advances not by sealing questions, but by learning to ask better ones. Each major leap in understanding has come not from certainty, but from the willingness to admit that something fundamental remained unresolved.

This is where the present discussion deliberately stops short of answers.

The aim has not been to replace one framework with another, but to reveal the deeper continuity beneath them. To show that what appear as competing explanations may, in fact, be partial views of a larger coherence. To suggest that beneath our models of force, space, and motion lies a more general structure—one that has been implicitly guiding our theories all along.

An open ending is not an abdication of responsibility. It is an acknowledgment that understanding grows through refinement, not closure. It respects the complexity of the universe by refusing to compress it prematurely into a final form.

What has been offered here is not a destination, but a direction. A way of seeing that aligns physical description with conceptual clarity. A reminder that science advances not only by solving problems, but by learning which questions deserve to be asked next.

The universe does not owe us final answers. It offers patterns, behaviors, and hints—enough to guide inquiry, not enough to conclude it. The role of understanding is not to dominate these patterns, but to remain responsive to them.

In this sense, the work does not end here. It opens outward.

The ideas presented are not meant to close debate but to refine it. They invite further exploration, reinterpretation, and testing. They ask the reader not to accept a new doctrine, but to adopt a new posture—one of attentiveness to structure, sensitivity to coherence, and humility before complexity.

If this work succeeds, it will not convince through finality but through clarity. It will leave the reader not with answers to memorize, but with questions sharpened by insight.

That is the true mark of understanding: not the certainty of having arrived, but the confidence to continue walking—aware, attentive, and open to what unfolds next.

Closing Reflection — Continuity Without Closure

Every serious inquiry eventually reaches a point where it must decide whether to end or to remain open. Not because the questions are exhausted, but because understanding has matured enough to recognize its own limits. This is such a moment.

What has unfolded through these chapters is not an argument in search of victory, nor a theory seeking replacement of others. It is an attempt to trace a continuity—between what has been explained and what still resists explanation; between the language we use and the reality it gestures toward; between the certainty of mathematics and the quiet ambiguity that always lies beneath it.

The journey has not been about disproving existing frameworks, but about recognizing their edges. Not to weaken them, but to see where they end. Every great structure of thought carries within it the outline of what it cannot contain. Recognizing that boundary is not failure—it is maturity.

What emerges, then, is not a final doctrine, but a refined stance. A way of holding knowledge lightly without diminishing its rigor. A way of honoring precision without mistaking it for completeness. The universe, as revealed through this lens, is not a closed system awaiting final explanation, but a coherent unfolding that invites ongoing understanding.

This perspective does not diminish science; it deepens it. It allows theory to remain alive—responsive rather than rigid. It restores humility to inquiry, not as a retreat from knowledge, but as an acknowledgment of its depth.

If there is a quiet message carried through these pages, it is this: understanding is not achieved by exhausting mystery, but by learning how to live with it responsibly. Meaning arises not from closure, but from continuity—from the willingness to keep asking without demanding finality.

The work, then, does not end here. It simply reaches a place of rest, a pause from which further insight can grow. The questions remain open, not because they are unanswered, but because they are alive.

And in that openness, the journey continues.

Author's Note

This book did not begin as a theory, nor was it written to challenge any established discipline. It began as a quiet discomfort—an awareness that while our explanations of the universe had grown remarkably precise, something essential remained unspoken.

Over the years, as I studied physics, philosophy, and the history of ideas, I noticed a recurring pattern. Each generation refined the language of understanding, yet the deeper questions remained unchanged. We became better at describing *how* the universe behaves, but increasingly hesitant to ask *what that behavior means*. Somewhere along the way, explanation began to substitute for understanding.

This book was written in that space of unease—not to overturn existing knowledge, but to sit beside it and ask what it quietly assumes. It does not claim to correct science or replace its achievements. Rather, it seeks to listen carefully to what those achievements imply when taken seriously and followed to their logical edge.

The ideas presented here are not conclusions but invitations. They do not demand agreement, only reflection. They are offered in the belief that inquiry deepens when it is allowed to remain open, and that clarity emerges not from certainty, but from sustained attention.

This work is not a rejection of physics, nor an attempt to redefine it. It is an effort to stand at its boundary—where description meets meaning—and to ask what lies just beyond.

If this book encourages the reader to pause, to question familiar assumptions, or to look again at ideas long taken for granted, it has served its purpose.

— *Sandeep J. Chavan*

Glossary

Alignment: The condition in which elements of a system operate in coherence with one another, allowing stability and meaningful interaction to emerge without external enforcement.

Boundary Condition: A constraint or limit that defines how a system behaves without specifying its internal mechanics. Often determines the range within which phenomena can occur.

Conceptual Framework: A structured way of interpreting observations. It shapes what questions are asked and what kinds of answers are considered valid.

Continuity: The principle that change occurs through connected progression rather than abrupt disjunction. In this work, continuity refers to the seamless relationship between structure, behavior, and interpretation.

Description: An account of how phenomena behave or appear, without necessarily explaining why they do so.

Equilibrium: A dynamic condition in which opposing tendencies balance each other, allowing stability without stasis. Not stillness, but sustained coherence.

Explanation: An attempt to account for underlying causes or principles that give rise to observable phenomena. Goes beyond description by addressing *why*.

Field: A conceptual structure representing how influence or interaction is distributed across space or conditions. Not necessarily a substance, but a pattern of relational possibility.

Framework: A system of ideas or assumptions that shapes interpretation. Frameworks guide understanding but do not define ultimate truth.

Geometry (Conceptual): A language for expressing relational structure rather than a literal description of physical substance. In this context, geometry is a mode of representation, not a material entity.

Interpretation: The act of assigning meaning to observation. Interpretation bridges data and understanding but is shaped by underlying assumptions.

Meaning: The coherence that arises when observations, concepts, and explanations align within a broader context. Meaning is not imposed but discovered through relationship.

Ontology: The study of what exists and in what manner. In this work, ontology is approached cautiously, as a question rather than a fixed answer.

Phenomenon: An observable event or pattern, regardless of its underlying cause.

Structure: The organized arrangement that allows phenomena to occur in a stable and intelligible way. Structure is relational rather than material.

Structured Vacuum Energy (SVE): A conceptual term referring to the organized, non-empty nature of the vacuum that allows interactions, propagation, and equilibrium. It denotes a functional substrate, not a physical substance.

System: A coherent set of relationships whose behavior arises from internal organization rather than isolated components.

Understanding: The integration of explanation and meaning. Understanding goes beyond description to grasp the significance of relationships within a system.

Universe: The totality of observable and inferable existence, approached here not as a static object but as an evolving relational structure.

Notes on Terminology

The language used throughout this book is deliberate, but not rigid. Many of the terms employed here—such as *structure*, *equilibrium*, *field*, *vacuum*, and *energy*—carry long histories across physics, philosophy, and everyday speech. In this work, they are not used to assert fixed definitions, but to guide understanding toward relationships rather than objects.

Where possible, familiar terms are retained to preserve continuity with existing discourse. However, their usage may differ subtly from conventional technical meanings. This is intentional. The goal is not to redefine terminology for its own sake, but to loosen it enough to allow deeper insight without forcing the reader into rigid theoretical commitments.

Some terms are used descriptively rather than formally. For example, words such as *structure*, *field*, or *equilibrium* are employed to indicate patterns of relationship and stability, not to assert specific mathematical constructs. Similarly, phrases like *vacuum*, *medium*, or *substrate* are used conceptually—to indicate underlying coherence—rather than to introduce new physical entities.

Where terminology overlaps with established scientific usage, the intent is not to challenge accepted definitions but to extend their interpretive range. This work does not propose new equations or predictive models; it offers a framework for re-examining assumptions that often remain implicit within technical language.

Readers are therefore encouraged to approach the terminology with flexibility. The meanings are contextual, evolving with the discussion, and meant to serve understanding rather than constrain it. Precision here is conceptual rather than formal—aimed at clarity of thought rather than mathematical exactness.

This approach reflects the spirit of the work itself: not to replace existing frameworks, but to look beneath them; not to redefine science, but to explore the foundations on which scientific thinking quietly rests.

In that sense, the terminology used throughout should be read as an invitation—an entry point into a deeper reflection on how we describe reality, and what those descriptions silently assume.

References

This work draws conceptually from a broad intellectual tradition rather than from a single discipline or school of thought. The following references represent foundational influences—historical, scientific, and philosophical—that have shaped the conceptual background of this book. They are not cited as authorities but as part of the ongoing conversation this work participates in.

1. Einstein, A. *Relativity: The Special and the General Theory.*
2. Einstein, A. *The Meaning of Relativity.*
3. Heisenberg, W. *Physics and Philosophy.*
4. Feynman, R. *The Character of Physical Law.*
5. Kuhn, T. S. *The Structure of Scientific Revolutions.*
6. Wheeler, J. A. *Geons, Black Holes, and Quantum Foam.*
7. Penrose, R. *The Road to Reality.*
8. Smolin, L. *Time Reborn.*
9. Rovelli, C. *Reality Is Not What It Seems.*
10. Bohm, D. *Wholeness and the Implicate Order.*

These works are not cited as authorities to be accepted, but as intellectual companions that helped frame the historical and philosophical landscape in which this book operates.

Further Reading (Forthcoming Works by the Author)

The ideas explored in this book are part of a broader, ongoing body of work examining the foundations of reality, knowledge, and human understanding. The following titles represent **planned and in-progress works** that extend the themes introduced here. They are listed to provide context and continuity, not as completed publications.

Each work explores a different dimension of the same central inquiry: how structure, meaning, and understanding arise within the universe.

- **The Still Universe:** An exploration of equilibrium, structure, and the limits of motion-based explanations in cosmology.
- **The Physics of Meaning:** A conceptual investigation into how meaning emerges from structure, interaction, and perception.
- **The Architecture of Reality:** A study of how order, coherence, and form arise without presupposing substance.
- **The Silent Geometry:** An examination of spatial relationships as expressions of structure rather than physical entities.
- **The Boundary of Knowing:** A philosophical inquiry into the limits of explanation and the nature of understanding.
- **The Language of the Cosmos:** A reflection on how

mathematical and conceptual languages shape our interpretation of the universe.

- **The Quiet Laws:** A study of stability, balance, and non-dominant principles governing natural systems.
- **The Question Beneath Physics:** An exploration of what remains when equations end and assumptions begin.

These forthcoming works are not sequels in a traditional sense. They are parallel explorations—each approaching the same foundational questions from a different perspective. Together, they form a long-term inquiry into how we understand reality, not by accumulating answers, but by refining the questions themselves.

Acknowledgments

This work would not exist without the long, often unspoken lineage of thinkers who chose inquiry over certainty. Though many of their names appear throughout the history of science and philosophy, their greatest contribution lies not in answers given, but in questions left open. This book stands on that tradition.

I am grateful to the scientists, philosophers, and thinkers—known and unknown—whose willingness to doubt, revise, and reflect made intellectual honesty possible. Their work formed the ground on which this exploration could take place.

I also acknowledge the quiet influence of those who question without seeking recognition: teachers who encourage thought rather than conformity, readers who engage deeply rather than quickly, and individuals who value understanding over conclusion. Their presence shapes the spirit in which this work was written.

Finally, I acknowledge the process itself—the slow, often uncertain movement from intuition to articulation. This book emerged not from certainty, but from sustained curiosity. If it carries any value, it is because it remains open to refinement, dialogue, and reinterpretation.

This work is offered not as an answer, but as an invitation to think more carefully, more patiently, and more freely about the nature of understanding itself.

About the Author

Sandeep J. Chavan is an independent thinker and writer whose work explores the deeper structure of reality beyond conventional disciplinary boundaries. His writing moves at the intersection of physics, philosophy, and conceptual inquiry, seeking not to replace existing knowledge but to question its foundations and assumptions.

Rather than approaching science as a collection of settled answers, he approaches it as an evolving dialogue—one that invites reflection on meaning, structure, and coherence. His work often examines where explanation ends and understanding begins, focusing on the spaces where established frameworks quietly reach their limits.

Through a series of interconnected writings, he explores how ideas such as energy, structure, causality, and consciousness may be understood not as isolated concepts but as expressions of a deeper, unified order. His approach favors clarity over complexity, insight over assertion, and curiosity over conclusion.

This book reflects that philosophy: not a declaration of final truth, but an invitation to think carefully, question freely, and remain open to what lies beyond familiar explanations.

Quote Bank

From What Einstein Did Exactly

A curated collection of reflective lines drawn from the spirit of the book. These may be used as epigraphs, section openers, or standalone quotations.

1. "Understanding does not begin where answers end, but where certainty pauses."
2. "A theory may describe the universe perfectly and still remain silent about what the universe is."
3. "The greatest misunderstanding in science is not error, but mistaking description for explanation."
4. "When a model works too well, it often hides the questions it was never meant to answer."
5. "We learned how the universe behaves long before we understood what it is."
6. "What we call reality is often just the limit of our language."
7. "Every equation carries an unspoken assumption; progress begins when we notice it."
8. "Geometry describes the path of motion, not the reason for motion."
9. "A theory becomes dangerous only when it stops inviting doubt."

10. "The universe did not become complex; our explanations did."
11. "To explain is not to conclude—it is to open a deeper question."
12. "What we call space may be nothing more than the shape of relation."
13. "Understanding begins where explanation ends."
14. "The greatest ideas endure not because they are complete, but because they remain open."
15. "When description replaces curiosity, knowledge stops growing."
16. "We mistake coherence for truth when we forget that coherence can be constructed."
17. "The deepest insights often appear not as answers, but as unease with existing ones."
18. "Reality does not resist explanation; it resists simplification."
19. "Every scientific revolution begins when someone asks whether the question itself is complete."
20. "To understand the universe is not to control it, but to listen to what it allows us to know."

Also by Sandeep Chavan

Foundations of Reality & Science

- *Reality Check: Form Is a Guest, Not a Host*
- *The Structure of Reality*
- *Unified Energy Dynamics*
- *Universal Energy Dynamics Canon (UED 2.0)*
- *Rethinking Energy: Ripple Tension, Not Substance*
- *Myth Wrapped in Math*

Physics, Meaning & Interpretation

- *Feeling Like a Physicist*
- *TIME – The New Definition*
- *$E \neq mc^2$: The Collapse of Classical Assumptions*
- *Physics Fixing*
- *The End of Particle Age*

Mind, Intelligence & Human Understanding

- *The Human Mind: Ripple of Awareness*
- *The Physics of Emotions*
- *Intelligence: It's Not What You Think*
- *Universal Skill: Navigating Consequences*

Philosophy, Ethics & Consequence

- *The Law of Consequences*
- *Ethics & Morale: Shades of Consequence*
- *The Law of Truth*
- *The Question That Won't Go Away*

Contemporary Reflections

- *Stop Trying to Be Happy*
- *Motivation Overrated*
- *What Collapses Your Reputation*
- *What I Never Told My Children*

Each of these works approaches the same central inquiry from a different angle: not what we know—but how we know, and what we quietly assume when we claim understanding.

Disclaimer

This book is a work of independent inquiry and philosophical reflection. It does not claim to present established scientific theory, empirical proof, or experimental validation in the conventional academic sense. The ideas expressed herein are interpretative, exploratory, and conceptual in nature, intended to stimulate thought rather than assert final authority.

While the work engages with concepts drawn from physics, cosmology, and philosophy, it does not seek to replace, revise, or contradict established scientific frameworks in a formal or technical capacity. Instead, it offers an interpretive lens—one that reflects on how knowledge is structured, how explanations are framed, and how meaning is constructed around scientific understanding.

Any references to scientific theories, models, or terminology are used illustratively and philosophically, not as claims of empirical revision or experimental proof. Readers are encouraged to treat the material as reflective inquiry rather than definitive scientific exposition.

The interpretations presented are the author's own and are offered in good faith as part of an ongoing exploration into the nature of understanding, not as authoritative conclusions.

Statement of Originality

This work is an original intellectual composition by the author. While it engages with ideas, theories, and perspectives found across physics, philosophy, and the history of science, all interpretations, syntheses, and conceptual framings presented here are the author's own.

No part of this book is intended to replicate, replace, or reinterpret any existing scientific theory in a formal or technical sense. Rather, it offers a distinct perspective that reflects the author's independent reasoning and long-term contemplation of foundational questions.

Any resemblance to existing theories or frameworks arises from shared engagement with common questions, not from duplication or derivation. The ideas presented are the result of personal inquiry, reflection, and synthesis, and are offered as an invitation to thought rather than a claim of authority.

This work stands as an original contribution in spirit and expression, intended to encourage dialogue, curiosity, and continued exploration rather than definitive conclusion.

Don't miss out!

Visit the website below and you can sign up to receive emails whenever Sandeep Chavan publishes a new book. There's no charge and no obligation.

https://books2read.com/r/B-A-EPGPC-QXDXI

BOOKS 2 READ

Connecting independent readers to independent writers.

Did you love *What Einstein Did Exactly*? Then you should read *$E \neq mc^2$: The Collapse of Classical Assumptions*[1] by Sandeep Chavan!

[2]

What if everything we thought we knew about energy is incomplete?

For over a century, $\mathbf{E = mc^2}$ has symbolized the elegance of physics—a compact equation that married mass with energy and reshaped the modern scientific worldview. But what if it also marked the beginning of a deeper misunderstanding?

In *$E \neq mc^2$: The Collapse of Classical Assumptions*, independent physicist and philosopher **Sandeep Chavan** presents a provocative new vision of reality—one where energy isn't stored in mass, light

1. https://books2read.com/u/4DaMLe

2. https://books2read.com/u/4DaMLe

isn't emitted from particles, and explosions are not releases, but realignments.

At the heart of this transformation is **Chavan's Realignment Equation**, a field-centric model that replaces classical concepts of mass, force, and emission with ripple-based coherence, structured vacuum dynamics, and a new limit on perception: the **Alignment Threshold Time (ATT)**. What Einstein treated as constants—mass, light speed, time—are reinterpreted as consequences of ripple synchronization within an intelligent field.

This book isn't just theoretical—it rewrites the practical language of energy, light, and decay. From **fire and nuclear bombs** to **the Big Bang** and **black holes**, each phenomenon is re-explored through the lens of **Universal Energy Dynamics (UED)**. Along the way, it challenges the foundations of particle physics, space-time models, and object-based reality.

Designed for scientists, engineers, philosophers, and curious thinkers alike, $E \neq mc^2$ blends rigorous insight with radical clarity. It features:

A full breakdown of the **Realignment Equation**Field-based alternatives to mass, inertia, and forceThe **Collapse Law**: why things vanish, not explodeImplications for measurement, engineering, and consciousnessA bold rethinking of **Einstein's legacy**—not discarded, but upgraded

This is not science fiction. It is a **post-Einstein doctrine** built on consequence, perception, and the hidden geometry of energy.

Whether you're a physicist looking for a new frontier, a student questioning the standard model, or a philosopher seeking a bridge between matter and meaning, this book opens a door into a universe that doesn't move—it **realigns**.

Nothing glows. Everything echoes. Nothing moves. Everything adjusts.

It's time to collapse old assumptions—and perceive energy as it truly is.

Read more at https://www.sandeepjchavan.com/.

www.ingramcontent.com/pod-product-compliance
Lightning Source LLC
LaVergne TN
LVHW100305210926
848480LV00003B/1030